FACTORY INFORMATION SYSTEMS

Design and Implementation
for CIM Management and Control

JOHN GAYLORD

Siemens Research and Technology Laboratories
Princeton, New Jersey

CRC Press
Taylor & Francis Group
Boca Raton London New York

CRC Press is an imprint of the
Taylor & Francis Group, an **informa** business

First published 1987 by Marcel Dekker, Inc.

Published 2019 by CRC Press
Taylor & Francis Group
6000 Broken Sound Parkway NW, Suite 300
Boca Raton, FL 33487-2742

© 1987 by Taylor & Francis Group, LLC
CRC Press is an imprint of Taylor & Francis Group, an Informa business

First issued in paperback 2019

No claim to original U.S. Government works

ISBN 13: 978-0-367-45145-5 (pbk)
ISBN 13: 978-0-8247-7389-2 (hbk)

Visit the Taylor & Francis Web site at
http://www.taylorandfrancis.com

and the CRC Press Web site at
http://www.crcpress.com

Library of Congress Cataloging-in-Publication Data

Gaylord, John [date]
    Factory information systems.

    (Manufacturing engineering and materials processing ;
23)
    Includes index.
    1. Computer integrated manufacturing systems—Design and construction. I. Title. II. Series.
TS155.8.G39   1987   670.42'7   87-9234
ISBN 0-8247-7389-6

Preface

A factory information system (FIS) is used to manage and control manufacturing operations. The FIS communicates with and connects the many other computer based systems used to support the factory. It is, therefore, the entity that binds the functions of process control, equipment maintenance, production scheduling, product accounting, and performance reporting into a consolidated interactive system for manufacturing control.

These computer based systems have been touted for many years as a means to improve industrial productivity and the quality of goods produced. Such systems do improve productivity and quality, but only if they are properly designed, applied, and supported by the correct incentives for their use. This book is written to help ensure that the FIS will meet these criteria. It is directed and dedicated to those people who, without much glamour and recognition, are involved in the preservation and enhancement of our historical manufacturing strength.

The book tells how to develop a successful factory information system. The approach is to show how an FIS improves productivity, how to define and implement a system, how to develop techniques to automatically detect production problems, and how to use a system to initiate actions to solve these problems. Typical situations that impede system development are described to help the reader develop strategies that avoid drawn-out implementation, excessive changes, inflexibility, and marginal usefulness. A guide is therefore provided through the morass of organizational, political, and economic problems that beset those who design and use these systems. Emerging technologies which are applicable to manufacturing — data communications, modular distributed computing, and artificial intelligence — are introduced. The result is a broadly integrated approach to the manufacturing systems field.

During my career in research, product design, and manufacturing it has become apparent that complex manufactured products do not move easily from design to production. Organizations that can rapidly accomplish this transition are acutely aware of their process capability and how to keep production "in control." Those who fail are trying to produce products without sufficient manufacturing control. This book gives approaches to this control problem, and will help those faced with the task of producing highly technical products.

Many people have contributed to this work. I am especially indebted to Dr. Karl Zaininger, President of Siemens Corporate Research Support, Inc., and Director of the Siemens Research and Technology Laboratories, and to Dr. Edward Devinney, Head of the Artificial Intelligence Department, for their encouragement and support. Earlier work at the RCA Technology Center and David Sarnoff Research Center involved many whose knowledge and experience proved extremely useful. Much was also learned from system users at numerous manufacturing sites in the United States, Mexico, and the Federal Republic of Germany.

I am therefore indebted to all those who have participated in the projects from which we learned this trade. Of special note is Mr. David Coleman, who dared to apply statistics to the real world of manufacturing control, and who contributed the fifth chapter of this book.

Those who gave the most, without visible reward, are my wife, family, and close friends, who encouraged throughout the entire writing process. To all these great people — a special thanks!

John Gaylord

Contents

1

Introduction to Computer Based Manufacturing Systems

1.1 INTRODUCTION

A factory is itself a society. It is composed of diverse people with a larger common goal but with differing and sometimes conflicting subgoals and incentives. The relationships between these people and the resources they control is very complex, as in any society. The speed and accuracy with which they communicate, resolve conflicts, and make decisions directly affects the productivity and efficiency of their manufacturing operation. This first chapter describes computer based systems that are used to assist in this social process of communication and decision making.

To function societies have evolved languages, communication systems, transportation techniques, and rules of behavior. Like production operations they have defined objectives and devised ways of measuring progress toward achieving these objectives. As

societies grew they made contact with other adjacent societies
which had different languages, ways of communicating, transportation, and rules of behavior. In much the same way, the factory
is in contact with its related organizations such as research centers,
product engineering functions, marketing operations, and financial
control groups, many of which are often remote from the manufacturing site.

Like isolated societies, manufacturing plants have also evolved
their own unique languages, ways of communication, and rules
of behavior which differ from product design or marketing societies. Thus when manufacturing engineers and product design engineers endeavor to move a new product into manufacturing these
differences can result in misinterpretations and loss of information
transfer. Another example is the difference in incentives and rules
of behavior for production control and marketing groups. These
differences lead to conflict between optimizing the manufacturing
operation and satisfying the customer demand. Conflicts must,
therefore, be resolved at relatively high levels both within the
manufacturing organization and between it and the other related
macrosocieties necessary to run the larger business.

For the larger business to succeed all the management and
control systems of the various macrosocieties should be connected
together. Within the factory the FIS system should be connected
to all levels of management including the production machines.
The concept of total optimization expressed by this interconnection is called Computer Integrated Manufacturing or CIM.

The objective of this first chapter is to describe in general
terms the various systems which comprise the composite CIM
system. It becomes readily apparent that as these systems interact
with each other, the same problems experienced by expanding
societies reoccur—languages are incoherent, communication systems are incompatible, and behavior is governed by different rules.
This, in summary, is today's CIM challenge.

1.2 COMPUTER INTEGRATED MANUFACTURING
(CIM) SUBSYSTEMS

Various computer based manufacturing systems shown in Figure
1.1 are used to improve the quality, speed, and cost effectiveness
with which new products are created and produced. When con-

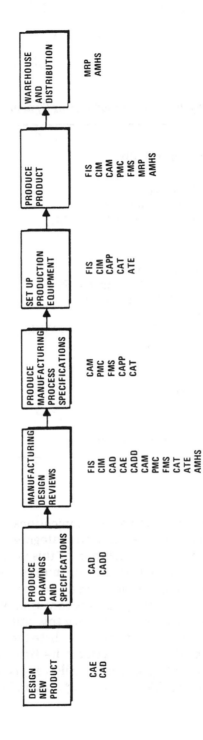

Figure 1.1   Computer based systems used for product evolution.

Table 1.1   Generic Computer Based Manufacturing Systems

---

1. *Systems Integration*
    Factory Information System (FIS)

2. *Design Support*
    Computer Aided Design (CAD)
        Computer Aided Engineering (CAE)
        Computer Aided Design and Drafting (CADD)
    Factory Planning

3. *Manufacturing Control*
    Computer Aided Manufacturing (CAM)
        Computer Assisted Process Planning (CAPP)
        Process Monitoring and Control (PMC)
        Flexible Manufacturing Systems (FMS)
        Automated Material Handling and Storage (AMHS)
        Distributed Monitoring and Control (DMC)
        Scheduling

4. *Testing*
    Computer Aided Testing (CAT)
        Automatic Test Equipment (ATE)

5. *Material Acquisition*
    Material Requirements Planning (MRP)

---

nected together with the effective communications, described in Chapter 7, these systems form a computer integrated manufacturing (CIM) network. The realization of this network is the current goal of many manufacturing companies because it is essential for their survival in an increasingly competitive world.

The scope is broad and the jargon is confusing when one is first exposed to these systems. To simplify this situation they have been reduced to the five generic types listed in Table 1.1. The functions performed by these systems are first described in summary. This is followed by a more detailed explanation of how they work together to provide decision support and control for the factory.

The first of these generic systems integrates many of the others into the CIM network. This factory information system (FIS) ties together design, manufacturing, test, and factory management systems, monitors overall manufacturing performance, and assists with local production floor inventory control and scheduling. It is the "backbone" of computer integrated manufacturing (CIM).

Design support systems such as computer aided design (CAD), computer aided engineering (CAE), computer aided design and drafting (CADD), and factory planning primarily support engineering, design, and drafting activities. They also significantly improve product manufacturability and directly provide many of the software "programs" used by CAM systems to guide and control production machines. The factory planning system is used by industrial engineers to plan new production facilities and to modernize old ones. The planning is accomplished by using a very sophisticated workstation which contains many tools for simulation, graphics, and financial analysis.

The control of manufacturing is accomplished using computer aided manufacturing (CAM), process monitoring and control (PMC), flexible manufacturing systems (FMS), computer assisted process planning (CAPP), automated material handling and storage (AMHS), product tracking and flow control (PT and FC), distributed monitoring and control (DMC), and scheduling systems. In addition to controlling manufacturing processes these systems simultaneously supply much of the data used by the factory information system (FIS) to monitor production performance. The CAPP system is used to define efficient groups of manufacturing processes by simulation using machine capacity data, human work capability factors, and optimum line balance techniques. A FMS system uses programmable robots, transfer devices, and processing machines to provide rapid change-over for assembling or forming many different products in small, intermixed batches. FMS makes possible a manufacturing strategy often referred to as "lot-size one." The AMHS system transports materials to and from workstations, maintains accurate inventories, and executes warehouse and distribution functions. The product tracking system monitors the location and state of product as it traverses the various steps of the manufacturing process. Flow control systems monitor the status of machines and dynamically alter the movement of product to different machines or process steps as conditions and schedules dictate. A scheduling system advises marketing, production control, supervisors, and forepersons regarding future product movement based on market requirements and the analysis of available resources.

Computer aided testing (CAT) and automatic test equipment (ATE) accomplish product adjustment and assure product quality. CAT systems, like CAM systems, also supply information to an FIS. Material acquisition is accomplished by using a material requirements planning (MRP) system tied closely to a scheduling system. It uses the FIS to monitor the status of order completion and material stocks. It generates a master schedule plan for material acquisition by using subsystems such as bill-of-material, inventory-transaction, scheduled receipts, shop-floor-control, capacity-requirements-planning, and purchasing.

The systems that comprise CIM are still in their infancy. As more powerful and cost effective computers evolve, these systems will be economically feasible to implement in small and medium-sized manufacturing plants as well as the larger installations where they now are used. Today the major impediment to CIM implementation is the difficulty of connecting these systems to each other. They have yet to be effectively integrated. One major reason has been the lack of a generic technical standard for communication. This standard, called the Manufacturing Automation Protocol or MAP, is rapidly being established by cooperating industrial and international organizations. By the late 1980s implementation of these standards using local area networks will occur permitting the "factory of the future" to be finally realized if data structures are adequately defined.

Integration has also been inhibited because some of the system functions cross traditional boundaries of classical organization responsibility. There is therefore a lack of middle management incentive to integrate. A third curtailment for integration is the unabashed exposure computer based manufacturing systems give to problems and performance deficiencies. This has, for example, significantly retarded the use of computer systems for monitoring and control of manufacturing in some communist countries where management requires greater flexibility in performance reporting. These systems are enthusiastically used only by confident groups of people who are comfortable with facts, flexibility, and change in pursuit of improved performance. If properly designed, integrated, and used, they have great potential for improving resource utilization as discussed in the next chapter. Taken together they assist in the evolution of new products through design, development, production, and distribution.

Each of the systems listed in Table 1.1 provide a primary function. They also fulfill the needs of other systems by exchanging with them considerable data and information. The FIS engineer must therefore be aware of the data requirements other systems may place upon his or her design and structure it accordingly. This means that early in any design both intra- and intersystem communications strategies must be established and the information required by each system defined. CIM can not evolve without this level of planning.

For a factory information system to be cost effective it must be useful for many years. This means the design has to permit a flexible evolution of capability with time. Unfortunately flexibility is not easily attained, especially with a classical hierarchical system as discussed in Section 1.3. Hierarchical systems are usually overly complex and therefore less responsive and more expensive than simpler single purpose configurations. The design of flexible systems also requires more engineering resources for two reasons. Many possible growth patterns have to be considered, which enlarges the scope of the design. Secondly, more system capacity also has to be included because the final scope of functions is not well defined. The practical constraints of time and money will limit objectives and the flexibility that can be accommodated. Attaining the optimum compromise between flexibility, resources, cost, and application scope is the challenge for a FIS designer. Because there are these compromises, the good system designer makes users aware of the trade-offs and shares with them from the onset of system design, the establishment of design objectives.

Table 1.2   Factory Information System Functions

Operations Analysis
  Data collection
  Data validation
  Data storage
  Data retrieval
  Through-put, yield, and quality analysis
  Statistical experimental design support
  Line balancing
  Methods analysis (labor and parts standards)
  Manufacturing economics
  Equipment maintenance
  Product flow simulation
  Vendor performance analysis
  Employee capability and training needs analysis
  Resource and capacity planning
Operations Control
  Floor inventory
  Work in process
  Starts scheduling
  Automatic problem detection
  Decision automation using artificial intelligence

The following sections provide greater detail about each of the major systems that are interconnected to provide factory control and management.

1.2.1   The Factory Information System (FIS)

The factory information system provides a computing facility with data bases to support the functions shown in Table 1.2. Because the FIS is the highest level of system under direct manufacturing control, it is the major tool used to manage production operations. It accomplishes this management by analyzing these operations and directing, by exception, supervisory attention to productivity, reliability, quality, product flow, and material flow problems before they become intractable. It also acts as the logical connector to CAM systems lower in the manufacturing hierarchy, to higher

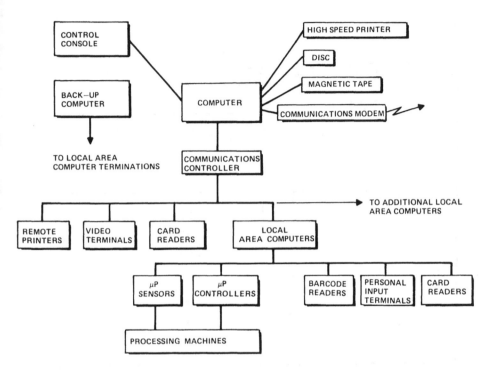

Figure 1.2 Typical large CAM system with FIS hierarchy.

level management information systems (MIS) which address financial control, marketing, and purchasing tasks, and to CAD systems for the receipt of process guidance programs for CAM. A FIS system can be hierarchical as shown in Figure 1.2 or distributed as shown in Figure 1.3.

The FIS functions shown in Table 1.2 differ from CAM functions in required degree of reliability and speed of response. This is because the FIS does not attempt to execute direct process control. The typical FIS system transaction occurs every few minutes compared to process control functions that occur almost continuously with subsecond reaction time. If the FIS system fails, production operations can usually continue with some manual support for a few hours as long as the process control systems are functioning. Thus, compared to CAM systems, the FIS system can

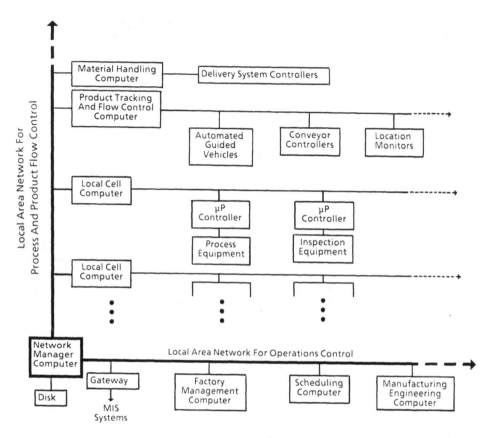

Figure 1.3   Typical distributed FIS.

be configured for slower response and without the redundancy
necessary to enhance reliability. This lowers its cost and is consistent with the need for reliable, real-time process control
functions in the controllers of processing equipment.

The FIS accomplishes operations analyses by using data that is
inherently collected by process controllers plus information that is
manually entered directly into the system. Some examples of these
data from controllers are production rates for each machine, reject
rates by machine and category, product splits determined by test

categorization, and production counts by machine. The FIS integrates data from each machine to provide a measure of department, activity, or cost center performance. Manually entered data are, reasons for machine down-time, maintenance performed, machine operator identification, visual inspection results, the production calendar, shift lengths, and so on. This information is not available from production machine controllers. It has to be entered into the FIS from other systems or humans via the video terminals, barcode readers, and mark-sense card readers.

The FIS data bases are the repository for the results of these analyses. Historical information regarding production capability and problems that affect productivity is thus available in the FIS system. The FIS is the ideal tool to evaluate new, more productive manufacturing technology because it contains within one system the necessary historical information on production problems and capability for contrast and justification.

1.2.2   The Computer Aided Design (CAD) System

As the name implies, this system supports new product design activities. It directly supplies engineers with the parametric information needed for design, a computational capability for processing this information, and an efficient means for laying out and drafting the results. The CAD system can be limited to drafting or expanded to include a data base and software programs applicable to complex design functions. The simplest CAD is an elegant digital drafting machine. The drawing is formed by generating a point on a video screen with a pen or tablet pointer followed by a series of instructions to produce the figure spatially related to the point. This is illustrated in Figure 1.4. The instruction capability includes mathematically defined figures and digitized shapes that are stored in memory accessible to the CAD system. Compared to conventional drafting techniques, figure generation, modification, and storage are very easy and rapid. This has made the purchase of these systems the easiest to justify by simply citing the increased drafting labor productivity.

Some CAD systems present figures in various color combinations to delineate specific surfaces or subassemblies as defined by the user. Many also provide a three-dimensional solid display or "wire frame," shown in Figure 1.5, to aid in visualizing complex

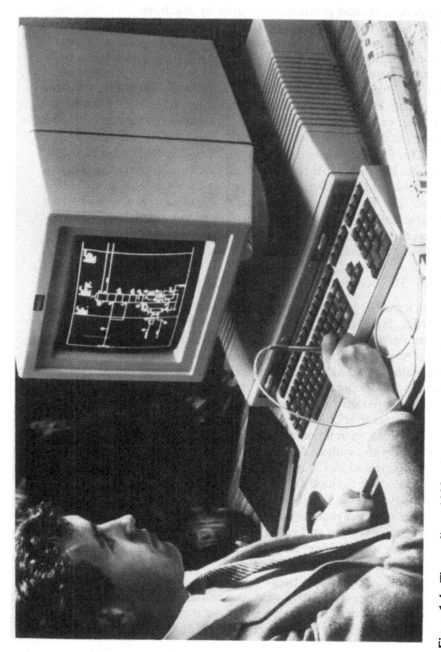

Figure 1.4 The applicon 4620 color raster-scan workstation (Courtesy of Applicon, Burlington, MA).

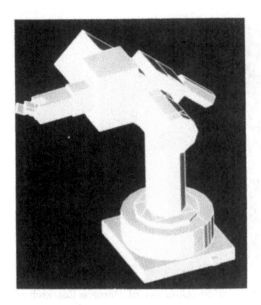

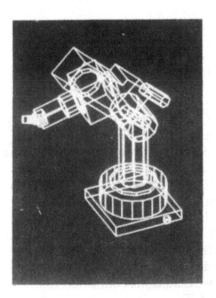

Figure 1.5 Examples of solid and wire frame images (Courtesy of Siemens Research and Technology Laboratory, Princeton, NJ).

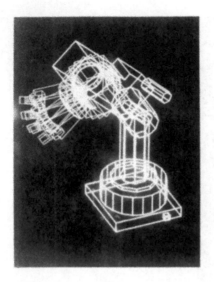

Figure 1.6 Example of motion simulation (Courtesy of Siemens Research and Technology Laboratory).

shapes. Associated software packages can display interference and clearance fits when parts are mated. A few of the more sophisticated systems even provide motion simulation, illustrated in Figure 1.6.

Today the most cost effective drafting systems are implemented using time-sharing. Time-sharing permits many user terminals to be operated from a single computer. This is possible because there are significant intervals when a draftsman or designer is not actively using computer resources. These unused resource intervals may be shared by queueing up computing tasks from many users thus improving system utilization. If too many terminals are used, the responsiveness of the system decreases. Each user must wait for the computer to process the tasks in the queue before it can respond to a request. About 3 seconds is the human tolerance limit for waiting. Multiple time-sharing workstations connected to a minicomputer are shown in Figure 1.7.

Data and figure transfer to other computer systems is by means of magnetic tape, disks, or transmission lines provides the means

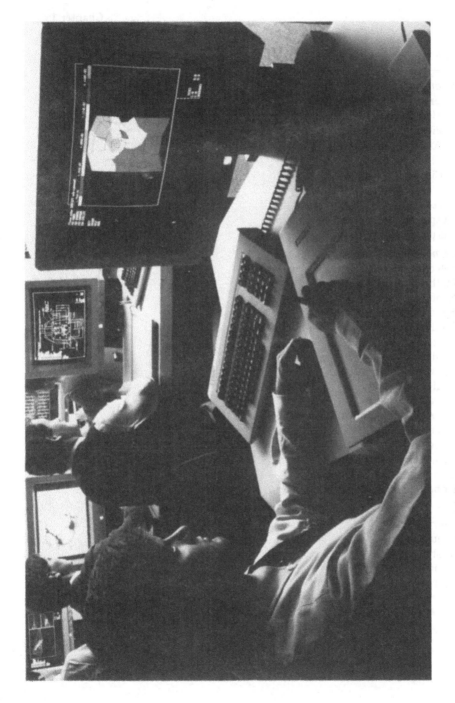

Figure 1.7   CAD/CAM time shared system using a DEC VAX computer (Courtesy of Applicon).

Table 1.3   Computer Aided Design Functions

Drafting
  Interactive graphics for
    Mechanics
    Piping
    Architectural drawing
    Flow charting
    Circuit schematics
    Circuit layout
    etc.
  Output for documentation control
  Output for operations control

Designer's Workbench
  Selection of components and materials
  Mathematics and statistics
  Heat flow analysis
  Fluid flow analysis
  Product flow analysis and simulation
  Cost analysis
  Analysis of fits
  Finite element analysis
  Logic simulation
  Circuit simulation
  Interactive component placement
  Circuit routing
  Design rule verification
  Connectivity extraction
  Test generation
  Robotic motion simulation
  etc.

Factory Planning
  Simulation
  Analytic modeling
  Financial analysis
  Drafting and layout graphics

---

for drafting results to be moved directly to production machines. This transferred information is then used to control manufacturing processes. The information typically guides such processes as machine tool selection and paths, photolithographic exposure patterns, time dependent digital thermal profile furnace controllers, automatic circuit board component insertion, robotic painting, welding, transfer operations, the sequencing of controllers, and many other production operations. This direct movement of digitized instead of written information from design to manufacturing is very cost effective because production set-up time and translation errors are reduced. In complex devices like very large scale integrated circuits the probability of making undetected human errors is large enough to make the transition from design to manufacturing impossible without CAD facilities. The various functions that a CAD system performs during the design of an integrated circuit are shown in Figure 1.8.

The more sophisticated CAD systems have, in addition to drafting and three-dimensional graphics capability, a battery of software programs to assist engineers with specific calculations and analyses. Some of these are listed in Table 1.3. The drafting workstation enhanced by access to these programs is better named a *workbench*. The concept of the engineers workbench is described by Dolotta et al. (1978). It began as the "programmers' workbench" used by software engineers at Bell Laboratories in 1973 to support corporate software development and maintenance. The typical workbench block diagram shown in Figure 1.9 brings to one location most of the resources needed by an engineer to accomplish a specific type of engineering. The efficiency of the engineer is thus improved along with the accuracy and quality of his or her work. The transfer of the design to manufacturing is more reliable and rapid, substantially improving this phase of product evolution.

The data base associated with a workbench initially contains parameters related only to the more theoretical and empirical aspects of the design function. These are used with appropriate analytic programs to accomplish the design tasks. As time passes the workbench matures by adding data that pertain to the pro-

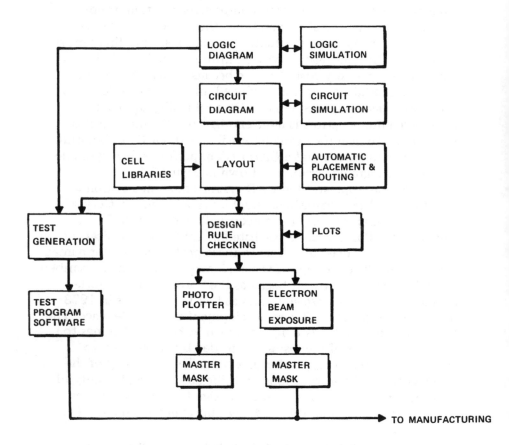

Figure 1.8   CAD mask generation and CAT test generation.

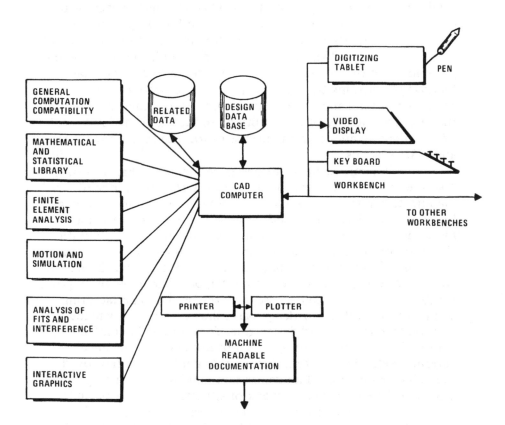

Figure 1.9   Typical computer aided engineering system.

duction problems and capabilities of the plant where the design is to be manufactured. This feedback of production dependent design limitations leads to improved design rules for manufacturability. The rules, which become endemic to design methodology by virture of being a part of the workbench data base, are then used for subsequent design guidance. Obvious benefit accrues because products that are engineered to avoid known production problems inherently are of better quality and can be produced at reduced cost. The CAD system therefore, in addition to supplying programs to direct manufacturing processes, also impacts manufacturing by making the basic design easier to produce. The reduction in time and labor required for new product introduction, the improved quality of the resulting product, and the lower cost of manufacture are powerful incentives for integrating the CAD and CAM systems.

The dynamic sequence of design refinements that improve manufacturability is illustrated in Figure 1.10. The source of manufacturing experience and technology is shown in the box at the top of the diagram. Situations arise in the course of manufacturing that uncover production problems and define limitations. The data base of the workbench is structured to accept this manufacturing information in an organized way for reference by design and manufacturing personnel. In the jargon of artificial intelligence this becomes a "knowledge base." As the design-manufacturing team work with this knowledge base, they generate new manufacturing technology and design rules for improved manufacturability. Better manufacturing methods immediately enhance current production performance. Design guidance via improved rules alter the next product generation making it easier to produce. The essential path is the feedback of situations, problems, and limitations from manufacturing to a repository in design. The essential action is the shared use of this knowledge base by both manufacturing and design personnel.

The shared use of a knowledge base to solve common problems is one illustration of the change that must occur if the full potential of these systems is to be realized. The technology to support remote access to distributed knowledge bases thousands of miles apart exists and is already being used extensively by other industries like banking and airlines.

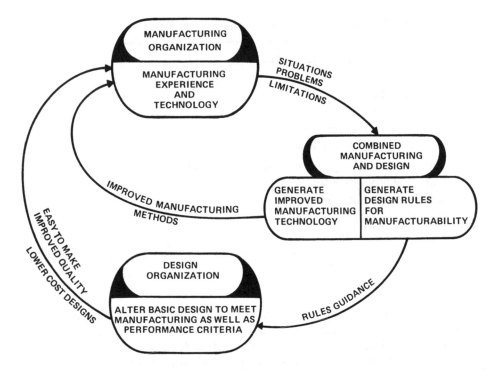

Figure 1.10 Manufacturing–design control loops.

1.2.3   The Computer Aided Manufacturing (CAM) System

The primary function of the CAM system is to control the manufacturing process. Examples of these control applications are
shown in Table 1.4. Programs resident in a CAM system direct
machine tools, instruct robots, and control processes. These
systems are thus in the mainstream of manufacturing. Other
systems can fail to operate for short periods of time without
drastic consequences. The CAM system must be very reliable and
in addition have well-defined manual methods of back-up for
the operations it controls.

The system size can vary from a single programmable controller to a large hierarchical or distributed group composed of a

Table 1.4   Computer Aided Manufacturing System Functions

Control Within a Manufacturing Cell

  Data collection

  Data validation

  Data retrieval

  Time sequenced process control

  Analytic model based process control

  Artificial intelligence model based process control

  Statistical process control

  Numerical control of machines

  Robotic control

  Product flow control

---

wide variety of sensors and controllers connected to one or more computers as was shown in Figures 1.2 and 1.3. The basic CAM data collection and manufacturing control function takes place within local areas of manufacturing called cells. There can be a number of these cells in a large factory system. Much of the hardware associated with a cell is often an integral part of the processing equipment supplied by an equipment vendor or custom built for the application. A well-designed system enables each cell to function alone for a period of time without support from the FIS. The period of independence depends on the anticipated time required to repair higher system levels if they fail, the data storage capacity within each cell, and the importance of data loss. The parts other than cells shown in Figures 1.2 and 1.3 are for system support, communication, data analysis, and report generation.

Many CAM systems evolve from small independent process controllers. As production facilities are added to a manufacturing operation the number of controllers grows. It eventually becomes obvious that a synergistic gain in total performance can be realized by connecting the controllers together. The hierarchical portion of the system is then added to affect this interconnection. If the control branches are truly independent, this hierarchical portion may not have to be extremely reliable. This reduction in required reliability permits the use of a more cost effective FIS to provide this interconnection as discussed in Section 1.2.1.

1.2.4   The Computer Aided Test (CAT) System

Computer aided test (CAT) systems use automatic test equipment (ATE) to do in-process, final, and quality assurance testing. The ATE systems use a computer to execute the test program. This software program sequentially applies a series of appropriate ambient conditions and test stimuli to the product by actuating hardware. After each application, an interval is programmed to control the effect of transient conditons. The test results are then sensed, collected, and stored before conditions are changed and the next stimuli is applied. The stored result, or past series of results, can thus be compared to some criteria and the result used to decide further action.

The entire test process is conducted under software control. The setting of test conditions, the timing of each test, the analysis of results, and the criteria to which the test results analysis is compared can thus be changed very efficiently. Some test applications, like the evaluation of large integrated circuits, have become so complex that computers are required to help write the test program software. These test generation systems derive the test conditions and expected results from the CAD output which defines circuit logic and layout. This is illustrated in Figure 1.8.

The most complex CAT systems automatically make adjustments to the product as part of the test procedure. In this case the manufacturing and testing processes become common. For example, in the chemical industry materials may be automatically dispensed into a solution based on the result of a test. After mixing the test is repeated and depending on the result more material may be added. This "adjustment" is repeated until the solution satisfies the limits of the test conditions. Other examples, as in automotive production, are the adjustment of a carburetor by servo driven tools during final engine assembly or the automatic adjustment of ignition circuit components to meet measured performance criteria.

Sophisticated CAT systems have become necessary to produce products that have many coincident interactive adjustments because we have reached the limits of human capability. A common example is our television set. Most modern producers of analog television receivers use a CAT system to dynamically monitor many set parameters and based on these, drive servos to adjust the many potentiometers and tuning coils to optimize set alignment. The result is improved color, focus, sound, and picture resolution. It is not possible for a human to do these many interdependent adjustments fast enough and with the consistency necessary for economical manufacture.

1.3   DISTRIBUTED VERSUS HIERARCHICAL SYSTEM CONFIGURATION

A factory is inherently an organization of spatially distributed work cells. Certain logically grouped production tasks are performed in each cell. Product is in an identifiable state as it enters each cell, is altered as it is processed by the cell, and leaves the cell again in an identifiable but now different state. Such a series of work cells forms a production system in which the tasks and the intelligence necessary for task execution is physically distributed throughout the factory. The smallest element or unit of intelligent task execution is a machine, person, or combination of both. These are the basic units that are combined within the cells to do the work required to change product states.

Until recently the structure of computing systems has not been a good match to the distributed nature of these factory operations. Problems of communication, data concurrency, and coincident processing were such that a central computer with its single task scheduler for control was the only economical and technically feasible system. Thus for years distributed factories have been served by hierarchical systems with a single computing facility at its head. This has caused a number of problems. Chief among these is a severe constraint on flexibility and evolutionary growth. A portion of a factory can and must be easily altered, enlarged, or deleted with minimal effect on the other operations. In the physical sense this can be done because factories are inherently modular and distributed. The software of a single computing facility can and should also be modular. Unfortunately when placed in a single hardware environment these software modules become functionally dependent on each other. Changes to the computing system corresponding to changes in the factory become difficult because of interdependencies within the computing system. The result is a slow degradation of system performance as the factory evolves and the computer system can not compliment the evolution. Symptomatically the system either can not be changed to perform certain desired new functions or it executes so slowly that its worth is reduced. Increasing computing power by adding a larger computer involves a major transfer of software and requires the solution of many hardware interface problems. Clustering of additional computers is not technically a sound approach because task allocation remains the basic problem, making the performance-to-cost ratio of clusters much less than for single machines.

The answer to this dilemma is to use the emerging technology of distributed computing systems. These systems, distributed like the factory, are composed of many small computers ("cells") which can communicate with each other such that they share their resources of memory, disks, printers, and so on, and do computing tasks for each other as well as execute their own programs. This is strikingly similar to how a factory works.

Each manufacturing cell works on its production task but, when needed the cells can help each other and share common resources. A distributed computer system in which each small computer serves the needs of a manufacturing cell is very compatible with the evolving needs of a factory. Modern local area networks discussed in Chapter 7 provide the "backbone" for this distributed factory system. Computing equipment can be added or removed from the communication network, in some cases while the system is operating, without affecting performance.

There are other advantages to the use of a distributed system in addition to its inherent flexible modularity. Such a system can be implemented one cell at a time provided the overall system concept is defined and the communication "backbone" is in place. This makes it possible to justify the cost of each small cell computer based on the cell's improved performance without having to make a costly up-front investment in a single large computer system. Properly designed distributed systems also degrade gracefully. This means the cell computer can fail or the communication backbone can fail and the other cell computers continue to function. Manual back-up procedures become more feasible. The lower cost of the small cell computers permit the stocking of spares for rapid substitution in the line when required. This alleviates many maintenance problems. Finally, the presence of a small computer at the work cell has a very positive psychological effect on operator morale. The operators can request *their* computer to report *their* progress during a shift. This sense of ownership of the cell including its computer and the immediate feedback of performance in the ever more technological and impersonal manufacturing environment can be a significant boost to the pleasure of work, with consequent increase in productivity as discussed by Northcraft (1985).

There are however some negative aspects to distributed systems. They involve issues of data integrity and task scheduling when data and processes are being executed concurrently on different machines. These problems have been solved however. A number of very reliable distributed workstations which are similar in function to computer controlled cells have proven their effectiveness and economy. Examples are the Apollo and Xerox stations. The consensus in 1987 of those involved in factory system design is to accept the remaining "bugs" of this new distributed technology in order to eliminate the well-established major weakness of the hierarchical center computer system.

A globally interconnected set of systems useful to a large manufacturing organization is shown in Figure 1.11. The realization of this manufacturing environment has been hindered by the cost of communications. Recently a rapid evolution in sophisticated communication technology has occurred to support the interconnection of these large systems. The amount of data and information available in an integrated system of this size far exceeds our ability to assimilate and use it without some form of automatic analysis. It is this data conversion to information and then the conversion of information to control directives that has to occur if these large systems are to provide their potential return on investment. Some techniques for accomplishing this auto-

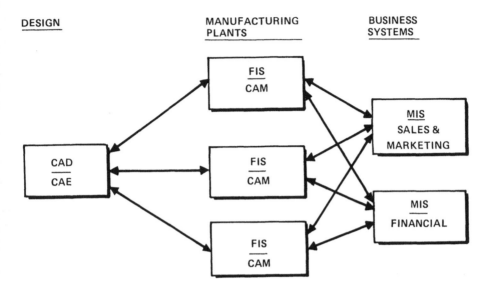

Figure 1.11   Global systems integration.

matic analysis and conversion to action are introduced in the next chapter.

REFERENCES

Dolotta, T. A., Haight, R. C., Mashey, J. R. (1978). "The Programmer's Workbench", *The Bell System Technical Journal,* July/August, *57*:(6), Part 2, p. 2177.

Northcraft, Gregory B. (November 1985). *A Managers Guide To Clean Room Operators*, Department of Management and Policy of The Center for Microcontamination Control, University of Arizona.

2

What to Monitor and Control: The Needs Analysis

2.1 INTRODUCTION

Chapter 1 introduced various computer-based systems that support manufacturing. This gave some insight into what these systems do and therefore how they assist with the manufacturing task. We now enhance this insight by a discussion of specific techniques for determining the needs of the manufacturing plant and for optimizing the use of manufacturing resources by using the Factory Information System (FIS) and those subsystems that communicate with it. The process of determining what to monitor and control to improve resource utilization is usually called a needs analysis. It is explained in this chapter and again reviewed from a different perspective in Chapter 8, which discusses the implementation of factory information systems. The needs analysis determines what is required to improve manufacturing performance. The scope of the needs analysis should be quite broad. If conducted in this manner, it usually results in the definition of organizational changes as well as functional requirements for the FIS. Once the needs are determined the FIS and associated systems can be designed to provide the functions required to meet these needs. Many of the functions involve an improvement in the way manufacturing resources are utilized.

2.2 RESOURCE UTILIZATION

The approach is to examine the various resources used by manufacturing, understand how they might be used more effectively, and, finally, define what the FIS can do to improve resource utilization. These resources are people, machines, and materials.

2.2.1 People Resources

People, our most complex resource, are probably used least efficiently. It is difficult to motivate nonprofessionals who work at the task of producing increasingly sophisticated, complex, and difficult to understand products. In contrast to inanimate machines, humans need to be motivated to function at their highest level of proficiency. This means the proper incentives, such as an understanding of the product being produced and recognition for excellent work, have to be in place. An awareness of educational needs and personnel capability must exist, and continuous training must be encouraged. Motivational incentives have proven to be very effective in both Western and Asian countries. In all cultures the primary motivation seems to come from the employee's identification of the relationship between his or her work and product quality and quantity. This is achieved through feedback to the worker on individual performance. Such feedback can be supplied by the FIS and should be one of the system design objectives.

Other examples of social and economic incentives are piecework, profit sharing, employee stock plans, and participation by all organizational levels in planning and decision making. The FIS system can help in the implementation of these motivational factors by measuring individual contribution fairly and without bias. The result is a more equitable allocation of both tangible and intangible rewards. A simple example is the accounting of piecework using weighting algorithms (that are clearly understood by the employees) for quality as well as quantity. Another example is the intangible stimulus to supervision from a comparison in real time of shift performance against standard rate.

The FIS can also assist with collective human decision making by clarifying the issues and by tracking the results of decisions after they are implemented to improve decision-making skills. The education of employees to improve skills can be directly supplemented by training programs resident in the FIS. If the FIS can monitor individual performance, it can also provide a measure of training needs. This type of feedback to employees has gained acceptance because it avoids personal discrimination when monitoring workers.

The author has experienced an example in which FIS monitoring of human performance surfaced a situation where the basic process was shown to be too cumbersome and complex for any person to control effectively. This resulted in the design of a better man-machine work cell, solving the problem and saving the operator from undeserved criticism. In this example, the plant management had lacked enough specific data to force this redesign prior to the implementation of the FIS.

National economies are such that today manufacturing is often located in depressed third world communities to take advantage of low labor rates and to supply these communities with financial aid. In many cases these workers have not had the opportunity for even moderate education. In addition, these people have little experience with the technically complex products being produced and the manufacturing systems used. In this situation, the FIS improves management effectiveness by providing a structured and disciplined way of extracting from the shop floor the information needed to manage a manufacturing operation populated by inexperienced workers. It also structures the flow of instructions to workers making the instructions consistent and concise, thus improving worker performance.

The computer-aided design system contributes to human resource utilization by saving time for those who design products. The hours required to execute a design are reduced because the CAD system rapidly provides most of the information and data needed for the design process. Documentation is faster because of

automatic drafting, text editors, and automatic parts list generation. This more complete documentation, achieved without dupli-
cating effort by engineers and drafting personnel, is easily and
reliably transferred to manufacturing. This eliminates more dupli-
cation of human effort and permits an earlier and steeper production learning curve. Because automatic design and manufacturing rule checking occurs, the designs are more reliable and
easier to manufacture making the production people, machines,
and material utilization more efficient.

The initial tasks for a new FIS are usually performance analysis and reporting. A well-designed system should also be capable of
providing additional functions. One example is resident programs
that are used to simulate production lines and product flow. The
use of these simulation programs increases the effectiveness of
manufacturing engineers in their utilization of floor space and their
balancing of production lines through methods optimization.

Administrative computer systems attached to the FIS can
also become the standards library for materials, parts, vendors, kit
list, methods, labor content, and cost. Because it is easy to maintain
these standards with word processing, less labor is required to support manufacturing in this administrative activity. As a library and
down-loading facility for software programs used to drive production equipment, the FIS reduces the labor and time lost from
mistakes in changing models and rearranging work flow. This will
become increasingly important as more robots and flexible programmable automation is added to our factories.

Test programs for automatic test equipment are also maintained and stored in these libraries with significantly less labor
required because of improved editing capability. Some FIS systems
regularly sample data from tests conducted on calibration
standards, analyze the calibration of the test equipment using
these data, and feed back corrections to the test control software.
In this way, the test facility resource is better utilized because
greater accuracy reduces the incidence of later retest and improves
yield by the reduction of guard bands on test limits. This could be
done manually but is usually not because of the extensive time

required and the necessary professional level of the individual conducting the analysis.

Finally, production managers can utilize their time more efficiently by using the FIS to get timely information for planning, decisions, and control. This occurs because data is automatically analyzed and converted to information. In addition, the information can be screened and reported only if certain management-by-exception conditions occur.

2.2.2   Machine Resources

Machines are the second resource available for effective utilzation. Improving machine utilization involves minimizing downtime, the time required for material transfer and processing, and the extension of the machine's useful life. This means effective equipment maintenance, production scheduling, and product flow control. To accomplish these goals, machines need to be smart and flexible. We are beginning to develop smart machines, able to learn from their environment and to make simple decisions. For example, they can change the flow of product through a cell to optimize machine utilization as discussed in Chapter 4. The capability for smart response is emerging from the computer science of artificial intelligence as discussed by Rich (1984).

Most machines can only exercise the rather rigid instructions programmed in their controllers. These instructions are not being changed dynamically as processing proceeds. The newer control programs take the form of "decision trees" where limits and data from sensors can be used to determine the path the program takes when it arrives at a logic "branch." Such machines are at least more flexible than "hard" automation because their dynamic control programs can be changed via software to produce new parts, assemblies, or models of product. FIS systems are often used to store and down-load these control programs directly to the production equipment as model changes occur. In return, the microprocessor-based controllers are a major source of data for the FIS, saving the cost of manually collecting performance information.

Historically, machines have been automated in groups as islands or cells of production to reduce labor. With the addition of mechanized product transfer between these cells, the automated factory can be connected together in two senses. First by the physical flow of product provided by these transport systems. The second is through the flow of information via the FIS. Considerable synergism occurs when information flow to and from production machines becomes available from the FIS. Management then has an overview of the machine group, permitting them to establish priorities for both product flow control and machine maintenance while concurrently measuring total production performance.

Automation requires a high level of integrated FIS communication and more sophisticated maintenance personnel. The emerging application of artificial intelligence to equipment maintenance is helping these personnel by simplifying the information communicated and by supplying rapid and more encompassing diagnostics. Without these apprentice or expert diagnostic systems based on artificial intelligence technology, it will be difficult for industry to utilize increasingly complex manufacturing technology.

2.2.3   Material Resources

Material resources are utilized effectively by eliminating waste through better product designs and more effective control of modern manufacturing processes. This results in smaller inventory, reduced scrap, less repair, and the reclamation of process by-products. The FIS contributes to all these functions by supplying better scheduling, control of factory operations, and up-to-date information regarding the status of materials on the production floor. This timely knowledge of the status of the factory is essential, for example, to support the just-in-time concept used to reduce material inventory.

New materials, such as reinforced plastics, that can be formed into parts on-line eliminate the scrap associated with machining and stamping operations. Parts inventories can be eliminated when these materials are molded into components as an integral part of the assembly operation. Regulations to control industrial pollution and the raising cost of basic materials make reclamation processing both mandatory and economical as well as ethical.

2.3 MANUFACTURING PARAMETERS TO MONITOR AND CONTROL

The objective of monitoring a factory is to rapidly sense the nature of the many operations, detect deviations from normal, and initiate appropriate corrective action. A thorough understanding of the manufacturing process is obviously required to define what parameters to monitor and what action to take. An analysis of what is needed to control the factory is based on this understanding. It is this needs analysis that defines the control functions required of the FIS.

The needs analysis usually results in a conclusion to which most production realists will readily agree: many manufacturing operations are not completely understood. This is because the manufacturing operations are always changing as new products are introduced and processes are refined. People often change their jobs, thus depleting the resident knowledge base of the manufacturing operation.

The strategy used to resolve this issue of process understanding and retention is twofold: (1) gain understanding by continuously learning more about how each process works and the relationship between processes and (2) maintain this knowledge by building it into the analysis and control programs executed by the FIS and associated process control systems.

2.3.1 Determining What to Monitor and Control

The determination of what to monitor and control so that the minimum amount of significant data can be collected and analyzed is achieved by using the iterative approach illustrated in Figure 2.1. Every factory, in fact, has been applying this technique since inception. The FIS contributes by supporting the monitoring, analysis, and control functions shown in Figure 2.1. This makes it easier to traverse the iterative cycle more rapidly. Coupling this with the FIS's ability to retain the knowledge gained as each cycle is traversed produces a rapid learning rate. This translates to faster start-up of new products and the optimum productivity for old products.

There are a number of useful techniques for systematically improving process understanding. These are (1) cause and effect diagrams, (2) evolutionary operation (EVOP) procedures, and (3)

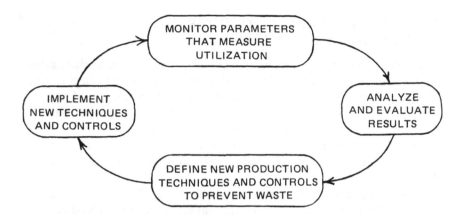

Figure 2.1   Improvement through iteration.

Pareto diagrams. These techniques, coupled with the methods used to structure the system design described in Chapter 3, are used to determine what to monitor and control.

Cause and Effect Diagrams

The cause and effect diagram was developed by Dr. Kaoru Ishikawa at the University of Tokyo in 1943 and is described in Ishikawa (1976), pp 18–28). It is a top-down diagrammatic technique used to help people determine the cause(s) for a given effect. This technique defines causes that can be monitored to give an early warning of trends which ultimately reduce productivity. To produce the diagram, a major arrow is drawn pointing to the effect. Minor arrows which generically categorize probable causes are then joined to the trunk of the main arrow. Subsets of arrows that define specific causes related to the general cause are then added. An abbreviated example is illustrated in Figure 2.2.

    In this simple example, the problem being addressed is the corrosion of automobile door panels. Corrosion is the effect of some set of causes. The general causes have been related to material, work methods, and production equipment. Specific causes are illustrated only for the materials category to simplify the diagram.

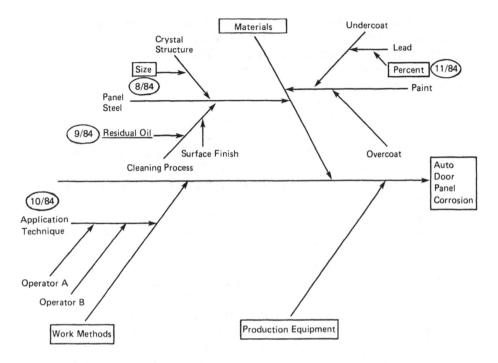

Figure 2.2   The cause and effect diagram.

These diagrams are helpful because they make people think in an organized and specific way about the reason for an effect. The underlying reasons for an effect become clear as the subordinate causes are defined. Finally, as the diagram is developed, measurable parameters appear at the ends of the branches. These become the parameters to monitor and control with an FIS system. Dr. Ishikawa suggests that these measurable parameters be placed in boxes on the diagram once they have been proven by some means to be actual and not just suspected causes. He also suggests underlining those parameters that are verified but can not be quantified. An example of this type of parameter is residual oil present after the cleaning process for the panel steel in the materials category.

The diagram can be improved by adding the date when the relationship between cause and effect was proven. The dates are shown in Figure 2.2 enclosed in circles. The cause and effect diagram thus shows both proven and hypothetical cause and effect relations and provides a chronological history of process understanding as it evolves.

The cause and effect diagram can also be constructed using the process flow diagram as the main trunk. This is appropriate to use when it is suspected that the effect being studied can be produced by a number of causes distributed through many process steps. Figure 2.3 illustrates this type of diagram for the corroded door panel example.

Occasionally, what appear to be independent causes are functionally related. When this happens, it is helpful to group the dependent causes at a higher level (closer to the trunk) of the diagram. One of the techniques used to establish and quantify these functional relationships is the scatter plot discussed by Moroney (1951). There is a cost associated with monitoring any parameter. To minimize this, only the independent parameters need be measured once the functional relationship is established.

Developing the cause and effect diagram is a technique to gain understanding. The technique works quite well during brain storming sessions where a number of people participate. It provides an excellent guide for discussion as other people are asked to help with the solution of a production problem. It achieves this by focusing the investigation. Once completed the cause and effect diagram is also very useful for informing others regarding the status of a problem or process.

Evolutionary Operation

Evolutionary operation (EVOP) is a term used to describe a special class of experimental designs applicable to ongoing production processes described by Box, Hunter, and Hunter (1978). The objective of EVOP is to optimize a process by producing statistically significant results from very small changes in process conditions. The changes are so small that they do not materially jeopardize the quality or quantity of goods produced. EVOP, therefore, can be a continuous effort associated with any pro-

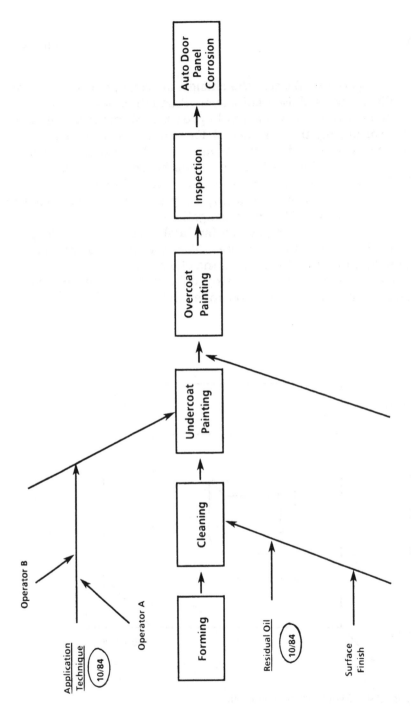

Figure 2.3 Cause and effect with total process.

duction operation. Because the changes are small, much data must be collected to establish significance. Production operations are applicable because they inherently produce enormous amounts of data. Fortunately the existence of Factory Information Systems with their sensors and computers has made the automatic collection of these data feasible. Before these tools became available, an EVOP investigation was very labor intensive and therefore expensive. Now, with an FIS system, it is practical to simultaneously conduct a number of EVOP investigations.

Usually, a simple replicated factorial design such as a 2exp2 or 2exp3 with center point is used. The concept is illustrated in Figure 2.4. In this example from the chemical industry, the volume of space shown by the large cube represents the extreme region boundaries for three controlled parameters of a process.

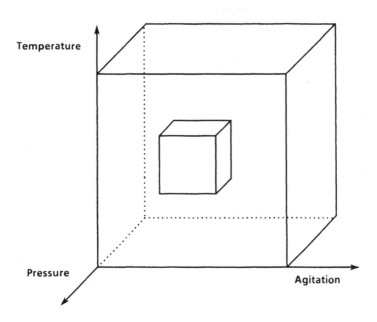

Figure 2.4   Evolutionary operation.

The controlled parameters for this 2exp3 example are agitation, temperature, and pressure. The objective of the investigation is to maximize the production of a plastic from this process.

In this scenario, the engineers and management of the plant want to achieve this goal but they are reluctant to make major changes in the three parameters because they cannot miss their short-term production commitment. They must, however, improve the process yield because they have run out of production capacity and face serious competitive pressure. Varying the process parameters to the extent represented by the boundaries of the large cube represents an unacceptable risk to immediate production output. The plant engineers are willing, however, to make small changes which are well within their experience with this process.

The first of these small changes is represented by the values of the controlled parameters at the corners of the small cube. A series of experiments are now begun, each typically running for a number of weeks. The initial experiment sets the controlled parameters at the center and at each corner of the small cube located at the middle of the current process range. By limiting the process variation to the boundaries of the small cube, negligible practical impact is made on quality and production. There is some very small effect, however, which when analyzed statistically becomes significant and shows that production would increase if the control parameters were changed in a particular direction. In our example, this would be toward the center of the small dotted cube shown in Figure 2.5.

To test this hypothesis a second experiment is then designed with controllable parameters set at the corners and center of this new cube. A few weeks later the results are analyzed and a third direction is indicated. This is subsequently tested with a third experiment. By continuing this evolutionary operation, the process is slowly adjusted until production is close to optimum.

It is beyond the scope of this text to explain in detail how these experiments are designed. This is discussed at length in Box, Hunter, and Hunter (1978). EVOP determines, for the particular manufacturing site, the relative importance of the various process parameters, their acceptable variation, and their closeness to optimum magnitude. This is exactly the kind of information required to select parameters for the FIS to monitor and control.

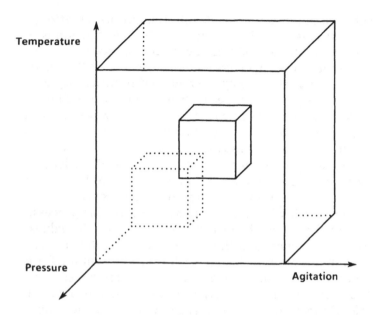

Figure 2.5   EVOP after first experiment.

Management-by-exception reports can then be generated for this
example process whenever any combination of the three parameters occurs that would reduce production by an arbitrary unacceptable amount. The important concept is that by monitoring
the correct control parameters, warning can occur prior to the
process moving out of control. Therefore management has time to
make corrections *before* a change in process adversely affects plant
performance.

Pareto Diagrams

The Pareto diagram discussed by Ishikawa (1976, pp. 42-50)
is used to visually quantify the relative influence various parameters
have on an effect (for example, productivity). The diagram was
conceived by Vilfredo Pareto, an engineer, sociologist, and eco-
nomist in the late 1800s. An example of his diagram is shown in

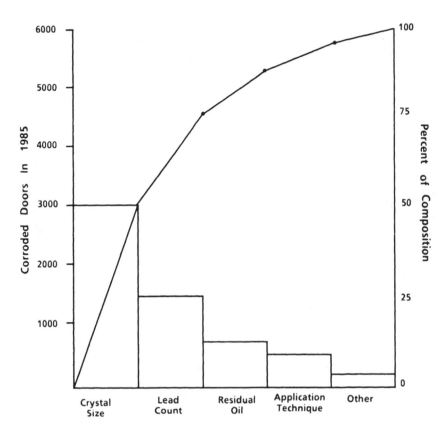

Figure 2.6   Pareto diagram.

Figure 2.6 and, for the sake of continuity, is related to the parameters used in the previous cause and effect diagram shown in
Figure 2.2. The diagram is plotted in descending order of impact
for the types of causes that produce the corrosion effect. The
vertical axes are usually shown in absolute units and relatively in
percent of the total. In our example, it is apparent that 75% of the
door panel corrosion problem can be attributed to the crystal
structure of the steel used for the doors and the lead in the paint.
The absolute scale is helpful when comparing the impact of a

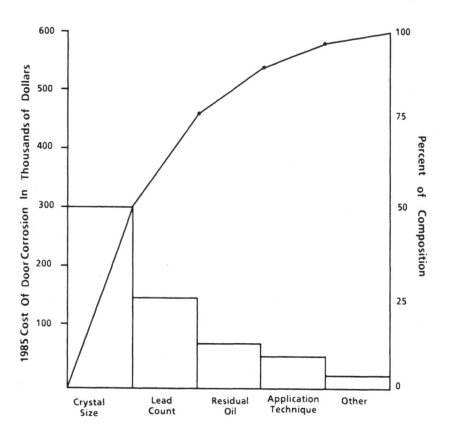

Figure 2.7   Pareto diagram in monetary units.

change in an effect as different parameters are optimized over a long period of time. It is also very effective to show the absolute scale in monetary units. In our example, if each corroded door costs the company one hundred dollars to repair, the Pareto diagram would appear as shown in Figure 2.7. This type of analysis reporting gets management's attention!

    One of the major hurdles facing those who design FIS systems is the communication of the results from the analysis of monitored data to a level of management that has little time, wants

quantitative and relative numbers, and controls the resources required to take appropriate corrective action. The Pareto diagram
is very helpful because it establishes the relative impact of various
causes graphically, therefore defining without lengthy discussion
the impact of taking different management actions. The FIS
design should, therefore, provide a graphics package with a simple
user interface so that the above diagrams can be generated easily
and quickly.

2.3.2   Basic Parameters That Measure
Resource Utilization

Earlier in this chapter the cause and effect (C-E) diagram was
discussed and it was suggested as a tool to assist in the definition
and documentation of parameters to monitor. The C-E diagram
shown in Figure 2.8 is now used to list the basic manufacturing
parameters that are classically measured and to show their relationships.

In most manufacturing operations there are two basic categories of parameters that are monitored. These reflect both the
history and status of (1) work-in-process and (2) processing equipment. The work-in-process is first identified as individual pieces or
as lots. The current position of this identified work is then monitored and as time passes this information is converted to past
movement history. The current status of the machines and people
doing the processing is also monitored. Again, as time passes these
data become the historical record of performance for the facility
or cell. Chapter 4 explains in greater detail how this monitoring of
product position and flow is achieved.

2.3.3   Accessing the Manufacturing Operation

In the mid-1980s, it is unusual for a person employed by a
manufacturing operation to initiate and define an FIS system.
More often people who understand computer systems, artificial
intelligence, applied statistics, and the technology associated
with designing the products being manufactured are asked to
conduct the needs analysis. They define the manufacturing parameters to be monitored, the actions to be initiated from these analyzes, and the FIS hardware and software configuration. The prob-

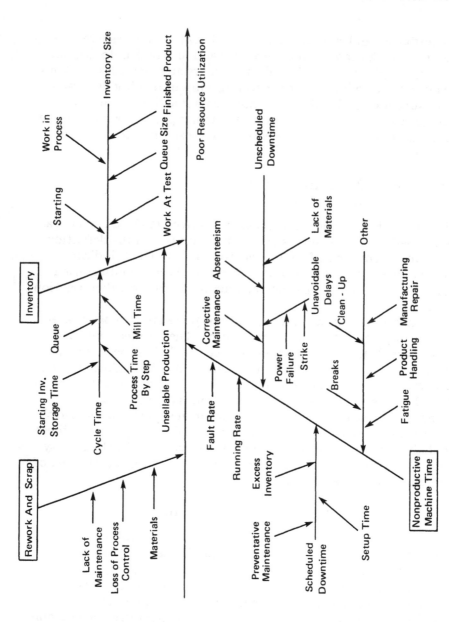

Figure 2.8   C-E diagram of basic parameters to measure.

lem with this approach is that people outside of the manufacturing organization are not privy to the internal drives of the operation and do not usually have an in-depth knowledge of the product or processes. They often define and produce technically successful systems that have less than optimum influence on plant operation and rather poor acceptance by the users. A desirable long-term strategy is to solve this problem by developing within the manufacturing organization the capability to do the needs analysis and an awareness of potential ways to meet these needs. Today, however, many small manufacturing operations cannot afford this type of professional staff. To achieve automation they must bring in outside consultants. To accomplish their job these consultants have to access the manufacturing operation with the support and participation of internal manufacturing personnel.

There are some approaches that these consultants can take to make this access more effective. Becoming a social as well as technical part of the manufacturing organization is necessary to understand and design the proper incentives into the FIS system. This means spending significant time at the factory. In addition to learning about the infrastructure of the organization, the time at the factory is spent obtaining five general types of information as discussed by Davis (1982). This information is used to define what will be monitored, the FIS data base structure, and the specific application programs for control and reporting. The five types of information are:

1. Identification of all things and events included within the scope of the FIS system
2. Relationships between the things and events
3. Attributes of the things and events
4. Validation criteria for the monitored data and for any parameters derived from the analysis of these data
5. How the data is to be analyzed and used by the FIS (required to define the data base structure and the specific application programs.)

There are four approaches to gaining this information. Most FISs are designed using a combination of the following:

1. Asking the manufacturing personnel
2. Derivation from existing data or system(s)
3. Synthesis from the basic needs of the manufacturing
    organization
4. Evolution

Asking

The most striaghtforward approach is to ask. This certainly
has to be exercised (1) to promote later use of the system and (2)
because production people are the only ones who have the required
detailed knowledge of what limits performance. There are, however, some problems with this approach. One is caused by the
human tendency of having better short-term memory than long-
term memory. This leads to overemphasis on the most recently
experienced problem as the most important need. There also is a
set of biases in any organization generated by current procedures,
available information, and inference from small samples. Innovation
can be limited by organizational policy, which was appropriate
in the past but should be changed in view of the new situation.
Finally, the roles played by individuals interviewed will affect
their responses. The interviewer must integrate these responses
sometimes without adequate knowledge of the operation.

Derivation from Existing Data

The needs can be derived from an existing manual system or
outmoded computer-based system. In the course of interviewing
users, examples of reports and data are always offered in abundant
quantity. The problem with this approach is that most old systems
also collect, analyze, and report slowly compared to the real time
interactive response capability of a modern FIS. Faster response
allows the control of manufacturing to be much more flexible
and effective. If the old transactions are simply copied by the new
system, these advantages are lost.

Old systems tend to perpetuate the collection and filing of
data that is no longer used. The needs analysis should identify
these data and eliminate their transactions and reports.

Synthesis

An approach that leads to fresh insight is the synthesis of the FIS applications from basic characteristics of the manufacturing operation. This is akin to starting over with the question. "Based on no previous knowledge or constraints how can this operation be better controlled?" This requires the system designer to spend extensive time at the manufacturing site to understand its objectives and how it operates. It is thus very laborious and time consuming.

Evolution

The last approach is to discover the needs by first designing a small, simple FIS which is easy to modify and then to support its evolution as it is used. The disadvantage of this evolutionary method is the very slow implementation rate which extends the time required to make an impact on manufacturing performance.

A combination of approaches should be used to determine what to monitor and control. The relative emphasis for each approach can be established by considering the ability of the users to specify their needs, and the ability of the system designer to elicit and analyze these needs. If the ability of both parties is low, the best approach is to start with a small FIS and let it evolve. With a little more combined ability, synthesis from a knowledge of the manufacturing operation should be emphasized. Still more ability permits derivation from existing data or an outmoded system. When both user and system designer are very able, asking is the most efficient approach.

2.4 MONITORING THE BASIC MANUFACTURING PARAMETERS EFFICIENTLY

The FIS designer must be careful to collect manufacturing data by utilizing production resources efficiently or the savings from using an FIS can be eroded by the cost of its data acquisition. This is especially true for the human resources. People still run the factories, and they are the only sources of some of the information required by an FIS. The cost of labor to enter this data manually can be very significant if the input process is slow. Manual data input can be slow because the entry process is cumbersome.

Table 2.1   Monitoring Nonproductive Time

Reported by people	Reported by machines

Scheduled down time because of:
   Preventive maintenance
   Excess inventory
   Set-up time
	Shift length
	
	Machine run time

Unscheduled down time caused by:
   Corrective maintenance
   Absenteeism
   Unavoidable delays
     power failure, and so on.
   Lack of material
	Total down time from:
	   Stopped time initiated by operator
	   Stopped time from process failure
	     Number of faults
	     Type of fault

Other causes:
   Operator breaks
   Clean-up
   Product repair during manufacturing
   Product handling
   Operator fatigue
	Processing rate

Using: Video data input terminal Marked card sense reader Voice input Digitizing tablet Dedicated key pad	*Using*: Direct data line to machine Punched paper tape (Old) Floppy disk Cassette tape Magnetic tape

Another problem with manually entered data is its inaccuracy. Humans make many errors which must be caught by validation as the data is input so that the data entry person, who is the only source for correction, can make the changes as part of the entry process.

The most cost effective and reliable method is automatic data collection directly from production and transport machines. In this section, we will review what kinds of basic data are currently input by people and machines and the types of data input equipment that is currently available for this function. These production data are usually classified by product type, model revision number,

Table 2.2 Monitoring Rework and Scrap

Reported by people	Reported by machines

Loss of process control caused by:
   Lack of maintenance
   Operator error
   Faulty material
   Wrong material
   Poor design
	Defect type producing faults
	
	 Loss of sensed process control
	
	 Portion of product reworked,
	     scrapped, or missing
	
	 History of reworked product

Using input devices cited in Table 2.1 plus:

Barcode reader
Magnetic card reader
Alterable memory attached to product
Production flow accounting system

and so on. For simplicity, these classifications are not shown in Table 2.2.

2.4.1 Monitoring Nonproductive Time

Table 2.1 summarizes by reporting category the data types that are collected and the facilities used.

2.4.2 Monitoring Rework and Scrap

Table 2.2 summarizes by reporting category some of the data types collected when monitoring rework and scrap. The devices used for data input are shown at the bottom of the table.

2.4.3 Monitoring Production and Inventory

Table 2.3 shows the data types collected or derived by an automated inventory and product flow control system.

    The effective control of the parameters listed in Tables 2.2 and 2.3 requires a product tracking and flow control system. This is discussed in Chapter 4.

Table 2.3   Monitoring Production Inventory

Starting inventory storage time	Real process time by step
Starting inventory size	Queue size of material handlers
Work in process	Finished product storage time
Work in test	Unsellable production
Finished product inventory	Cycle or flow through time
	Production rate and quantity by step

Using the same devices cited in Tables 2.1 and 2.2 plus:

Automatic counting and weighing

2.5   DATA ANALYSIS AND FACTORY CONTROL

The primary objective of an FIS should be to control rather than to report because only the control results in tangible performance improvement. The general approach is to use a control loop like the one shown in Figure 2.9. At the top of this loop,

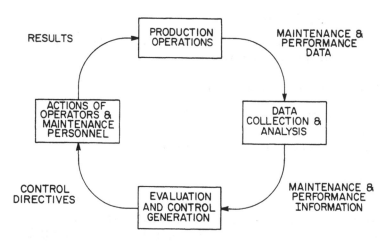

Figure 2.9   Data-information-control-action loop.

production operations produce a steady stream of maintenance and performance data related to the factory operations. These data get collected and analyzed by the FIS. The analysis converts data into information related to the maintenance problems and the quality of performance. This information is now in a management-by-exception form that is easy for management, supervisors, and operators to evaluate. After evaluation, decisions are made and control directives are generated. The control directives are transmitted to line operators and maintenance personnel where they are converted to actions which produce results that affect the production operations.

2.5.1   Control Loops

The generic control system (Figure 2.9) uses humans to evaluate information and generate control directives. Open loop control is slower and less reliable because humans remain in the loop. Today, however, it is the only alternative to implementing a FIS because the control algorithms are very complex and heuristic in nature. Artificial intelligence (AI) is being increasingly used to improve the responsiveness and reliability of this evaluation, and control generation function as discussed in Chapter 6. Unfortunately AI is not yet capable of dealing with the complexity of higher level decisions. This will occur, however in the next decade.

The more the functions of data analysis, evaluation, and control generation can be done by computer, the faster and more consistent will be the control. Large closed loop systems requiring fast reaction for control depend solely on computers for this function. For example, the launch of the first series of U.S. space shuttles was turned over to computers 36 minutes before take-off. In one instance a launch was aborted during the last 3 seconds of the countdown because during ignition, one of the engines stopped burning when a valve malfunctioned. This rapid data collection from the engine and valve, the interpretation of the data with the conclusion that the engine had stopped because of the valve and therefore it was safe to shut down after another engine had started, saved the ship and its astronauts from possible destruction. This level of decision automation can be applied to manufacturing today by using the FIS hardware listed in Section

2.3 and more sophisticated software. Implementation depends on economics because the technology exists.

Beside control there are administrative functions worth incorporating in the FIS. These functions are the anaylsis and reporting of performance in terms of yield, production, and quality, the maintenance of a standards library; and, the introduction of consistency in the way operations are controlled and reported. In a sense, these are still control loops but very open in nature.

The scope of FIS data analysis and control is quite broad. It can range from a specific tool in a head of a particular machine to the controller of the machine which may also control other machines, to the department, area in the factory, plant location, and division of the company. Periods over which data are analyzed can also vary from seconds to years. Because the scope of the applications and their range in time is so large, the system designer must constrain and structure these control loops with well-defined boundaries. The limitations and what the applications will not do have to be understood by the users before implementation begins or disappointment and dissatisfaction will result.

2.5.2 Communication Levels

The transfer of information from the FIS to humans can be done at two levels. The simplest is in directives for decisive action. To communicate at this level, the FIS must incorporate decision automation to produce the control directives. This technique is evolving as forms of statistical analyses and artificial intelligence are incorporated. The advantage of decision automation is the discipline it brings to the manufacturing operation. This is particularly valuable if the operators and first level of supervision are inexperienced. Unfortunately, decision automation stymies creative thought. The system will not evolve to keep up with changing products and technology if decision making is embedded in the software instead of the minds of the users. This is why a second level must exist. At this second level, factual information is transferred but the entire decision process is left to the user. There is therefore a role for both levels of information transfer in any system. The best allocation has to be judged for each individual site.

2.5.3   Initiation of Analysis and Control

Some incident has to start the generation of each analysis and
control action. This incident can be a request from a human, a
point in time programmed for the computers' clock, some limit
being exceeded, a heuristic-based decision from an AI program,
or the result of a complex multivariable analysis. Investigations
of performance problems are the usual reason for requests for
information by humans. Some analyses are driven by time, like
the end of the month status report. The control of machines,
processes, and maintenance is more often initiated by complex
analyses.

There are three basic situations that can trigger a response
based on comparing conditions against limits. The simplest situation is to test a single variable against fixed limits. An example is
when pressure exceeds the fixed safe limit, a valve is opened automatically to provide relief.

In the second situation many variables may be tested at once
and only if certain combinations occur will action be taken. For
examlple, the reject rate for a machine may exceed a limit but
other conditions affect the decision to perform maintenance. If
the machine is not used very often compared to other machines
that run continuously, it is better to maintain those that are
heavily used. This is particularly true if the maintenance staff is
limited. In this case the test is against a multivariable limit such
as the product of reject rate and run time factored by a staffing
condition. The third situation makes use of statistically based
control charts and AI programs to evaluate non-quantifiable
conditions. These complex triggers provide a way of initiating
many actions for process and management control. They are discussed in Chapters 5 and 6.

Any analyses and controls that are initiated by a set of prede-
termined conditions are management-by-exception driven. One
of the great services an FIS provedes for the plant management is
this automatic detection and reporting of problems or performance
that needs attention. This capability saves them a great deal of time
and helps to establish effective priority for action.

REFERENCES

Box, George, Hunter, J. Stuart, and Hunter, William G. (1978). Statistics for Experimenters. Wiley, New York, pp. 362-368.

Davis, G. B. (1982). Strategies for information requirements determination, *IBM Systems Journal*, *21*:(1), pp. 4-30.

Ishikawa, Kaoru (1976). *Guide to Quality Control*, Asian Productivity Organization, Tokyo, Japan. (obtained from American Society for Quality Control, Milwaukee, Wis.) pp. 18-28.

Moroney, M. J. (1951). *Facts from Figures*. Penguin Books, New York, pp. 271-320.

Rich, Elaine (1984). The gradual expansion of artificial intelligence, *Computer* (IEEE) *17*:(5), May pp. 4-12.

3

Structured Systems Analysis

3.1 INTRODUCTION

This chapter describes a methodology of structured design for Factory Information Systems (FIS) and other computerbased systems used in manufacturing. The methodology was adapted from techniques for the design of software systems described by DeMarco (1978, 1979). Structured design is important because it provides four major benefits to the FIS designer and user. These are summarized below.

1. How the system is used becomes an integral part of the design process.
2. The quality of the system is improved because a better needs analysis occurs.

3. The effort expended to specify the system is reduced.
4. Documentation for training system users is produced as part of the design process.

3.2 THE FLOW OF PRODUCT, CONTROL, AND DATA

A Factory Information System (FIS) is used to collect data, convert it to information, and communicate the information to people so they can use it to make decisions. The system also assists in changing the decisions into actions that improve productivity by communicating the control directives, derived from the decisions, to appropriate people and machines whose location and tasks are in turn determined by the location of product. The design of a FIS is thus concerned with the flow of data, information, control directives, and product.

The easiest part to understand is the portion of the design concerned with the flow of product because this is physically observable. Product flow refers to the movement of materials, or assemblies being produced, along well-defined physical paths. For continuously processed materials, these paths are pipes, reactors, storage tanks, and so on. Conveyors, moving belts or hangers, transfer machines, automatic assembly machines, and transfer robots are some of the paths for assembled products. Product flow diagrams are essential for plant layout, work-in-process control, and scheduling. They describe the physical plant and form the basis for modeling the dynamics of product movement. Chapter 4 describes these diagrams in detail and discussess how they are used in the design of systems for product tracking and movement or flow control.

In contrast to product, the flow of control through a factory is less obvious because it is not physically apparent. It is the movement of verbal or written directives which originate at some authority and pass through a chain of command to result in a directed action. Control flow begins with decisions and ends with actions. The actions usually directly impact product flow. The decisions usually originate with information derived from the analysis of data.

Data flow is the movement of data through the process of their collection, verification, analysis, conversion to information, and presentation in an easily understood form to authority. Data

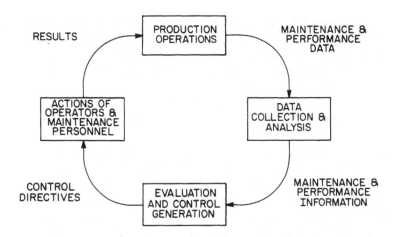

Figure 3.1   The data-information-control-action loop.

therefore can have two meanings. It can be the raw parametric or
attribute numbers which are obtained from some source, or the
results of the analysis of raw data. This latter form of data is usually
called information. Data have always been collected to measure
the performance of manufacturing operations. Now, more accurate
and larger quantities of data can be collected by the use of sensors
and computers. The computers are also used to do much of the
data analysis function and to present the derived information in
graphical form or summarized in management-by-exception reports.

The relationship between the flow of product, control, and
data is illustrated in Figure 3.1. Motion around this flow loop is
continuous. The objective of all this motion is to learn from each
decision how to make the next one better. The shorter the cycle
time around the loop, the more chances there are to suceed. The
decisions being tested usually come from many people at various
levels of the manufacturing organization. The loop thus provides
a way of evaluating many decisions simultaneously and studying
the interaction between these decisions.

Structured design encompasses more than just the data and
information portion of the flow loop. Control flow and the re-

sulting actions taken with the manufacturing process are included as part of the design procedure. FIS users and designers participate in defining how they will evaluate each type of information, what the resulting directives will be, and who will be responsible for converting these directives to action.

Because of computers, data collection and analysis have become much more economically feasible and therefore the quantity of available manufacturing data has increased. To maintain and improve the quality of these data, better planning to determine the true manufacturing data needs and how to meet them is required. Data, information, and control flow diagramming—along with the supporting methodologies of mini-specifications, data dictionaries, and mock-ups—are tools for manufacturing people to use in accomplishing this better planning of realistic manufacturing systems.

3.3   DATA-FLOW DIAGRAMMING

A data-flow diagram shows data needed for analysis and decision making, sources of the data, what the analysis is, and what is done with the information produced by the analysis. The diagramming technique greatly simplifies manufacturing needs analysis, system design, and eventual system use by providing a common graphical document that people with widely varied disciplines can easily interpret and understand. The data-flow diagram is a set of symbols representing sources and sinks for data, actions taken using the data, data storage files, and documents produced. Vectors are used to tell where and what data is transferred between the symbols.

The symbols shown in Figure 3.2 are suggested for manufacturing systems. It is essential for clear and concise communi-

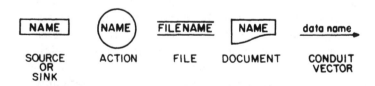

Figure 3.2   Data flow diagram symbols.

cation that the same symbols be used by the many people from different organizations who usually work on a manufacturing system project.

A *source* symbol represents the origin of data. The origin can be a person, machine, or any tangible physical thing that produces data. The *sink* is a repository for data or for the information derived from the data. Usually the sink is a machine or person that needs the data to make decisions and exercise control. Sources and sinks are terminators of the data-flow diagram in the sense of initiating the flow and receiving the results. The *action* symbol defines what is done with the data to transform it into more useful information. Actions are identified by a hierarchical numbering scheme that will be described later in more detail. Data storage is represented by the *files* symbol. The file symbol can have an optional header to ease diagramming. These files are physically located in computer storage devices (disks, memory, magnetic tapes, and so on). *Documents* can also be used to store data but more often are used to transfer information to a person, who can be thought of as a sink for some action. The *conduit vector* indicates the path and direction of data flow. Its label defines the specific data being conducted.

To illustrate data-flow diagramming consider the task of defining and documenting what a hypothetical production system such as the familiar popcorn popper should do. The verb *do* implies that the production system will take a series of actions to accomplish specific tasks. The first specific task is to test for out-of-control conditions in the popping process. The action symbol is shown in Figure 3.3.

Figure 3.3   Action symbol.

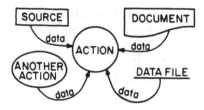

Figure 3.4   Generalized sources of data.

The next step is to determine what data the system needs to accomplish the action. This is always some kind of data from the generalized sources shown in Figure 3.4. For the popper control system the specific sources are shown in Figure 3.5.

The data-flow diagram also shows where the data or information resulting from the action are used. The general case is shown in Figure 3.6. The popper control system with repositories added is shown in Figure 3.7.

The symbols of data-flow diagrams are connected together to represent a series of actions. For example, the test popper control action can be considered as three separate actions. The first is the collection of temperature data. The second is the processing of these data along with the control limits to establish whether the process is in control. The third is the plotting of a control chart. The data-flow diagram of Figure 3.7 is expanded to show this greater detail in Figure 3.8.

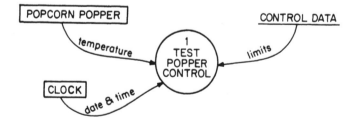

Figure 3.5   Popper control system data sources.

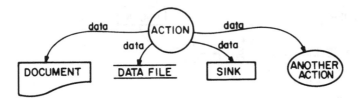

Figure 3.6   General repositories of data.

Data-flow diagrams can thus be expanded by breaking down
actions into greater detail. Conversely, they can be contracted
into simpler, less detailed representations by combining multiple
actions into a single action. The choice depends on the complexity
of the action and the level of detail desired. The rule of thumb
is to keep each diagram size small enough to fit on an 8 X 11-inch
page.

Most systems require a number of diagram levels for adequate
definition. The top level uses actions which are very broad in
scope. These actions are represented in greater and greater detail
in the underlying diagram levels. The diagram levels are designated
by a hierarchical numbering system. For example the first level of
the "Test Popper Control" action of Figure 3.7 is identified with
the number "1." In Figure 3.8 this same action is divided into
three actions. They are identified as 1.1, 1.2, and 1.3.

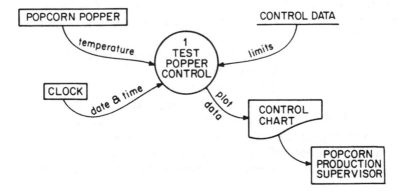

Figure 3.7   Popper control system with data repositories.

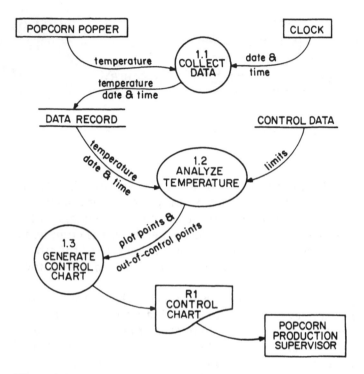

Figure 3.8   Detailed popper control system data flow.

If a shift occurs in the average measured temperature, the
temperature setting of the popper controller has to be adjusted.
The first level diagram of Figure 3.7 would then change to include
the control loop as shown in Figure 3.9.

The more detailed data-flow diagram describing the "Reset
Thermostat" action would be comprised of a series of detailed
actions labeled 2.1, 2.2, 2.3, and so on.

If this were done, the popcorn popper system would be described by a single first level and two second level diagrams, with
all of them fitting on three 8 × 11-inch pages. These pages, like
the contents of a good reference book, provide both a broadly
scoped and a detailed description of what is to happen. In addition

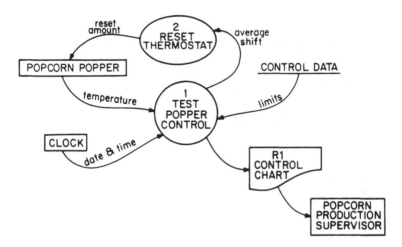

Figure 3.9   Corn popper data flow.

they lead, like an index, to the necessary supporting documents which completely define the system. These documents are the mini-specification, data dictionary, and mock-ups to be described later in this chapter.

The second level actions can be further detailed by creating a third level, (for example, 1.1.1, 1.1.2, 1.1.3, ...) and so forth. In this way any action can be divided into greater and greater detail until it reaches the level necessary for system implementation. The action symbols are labeled hierarchically as illustrated so the documentation can be easily organized.

A powerful aspect of this methodology is its modularity. An action symbol represents a module. It can be changed without affecting the other modules, and therefore without affecting the total system design. This independence occurs because each action module at any level is a black box with certain inputs and outputs. As long as these inputs and outputs remain the same, the system does not care what goes on in detail within the action module.

This modularity can be represented on the data-flow diagram by boundaries around an action or groups of actions. For example, if the action 1.2 in Figure 3.8 is found to use computer resources

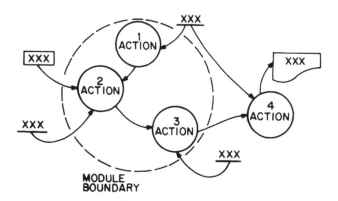

Figure 3.10   Modularity of an area of actions.

very poorly, thus slowing down the responsiveness of the popcorn manufacturing system, it can be changed by perhaps making software or hardware modifications. This can occur without the propagation of changes through the rest of the system because the input across the boundary (from the data record and control data files) and the output across the boundary (to generate the control chart) remain the same. Thus, to preserve the advantage of modularity the form and content of whatever crosses the boundary must remain invariant.

We have seen that actions can be combined or divided. This leads, a priori, to the conclusion that the module boundary can encompass a number of actions. This is illustrated by actions 1, 2, and 3 in Figure 3.10.

The bounded area can be considered as a module which can be changed independently as long as the type of data transferred by the vectors crossing the boundary does not change. The boundary can define a physical area where the described actions take place. For instance it may represent a specific machine, department, or building of the plant complex. Systems designed using structured methodology can automatically provide the desired independence between organizations by using the boundary con-

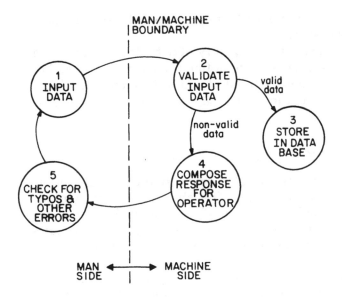

Figure 3.11   The man-machine boundary.

cept and simultaneously tie the organizations together through the data flow paths (conduit vectors).

The boundary is also a very convenient way to specifically define the interface between people, machines, and organizations. This helps to clarify system user responsibilities during needs analysis and later during the start-up of production use. The use of boundaries to separate man and machines is illustrated in Figure 3.11. The area to the right of the boundary is the computer system. The area to the left is the human environment. Data that cross the boundary into the computer must originate with people. Data from the computer, that is intended for people, cross the boundary in the opposite direction. To design and optimize the human interface of a computer-based system, one only has to address the problems defined by the data flow across the boundary. This greatly simplifies the interface problem by specifically defining what the issues are and where they occur. When the system has multiple users, in different locations and at different

organizational levels, the boundary technique helps communicate
and clarify these issues.

Modularity is an intrinsic quality of data-flow diagramming.
If data-flow diagrams are used to define what a system should do
and the system is so implemented, it will be inherently modular.
This means the system can more easily evolve as needs change with
time.

3.4 MINI-SPECIFICATIONS

The mini-specification is a detailed description of a single action
described in a data-flow diagram. The label of an action symbol
suffices for the broadly scoped data-flow diagram, but is not detailed enough to adequately define how the action is to be carried
out. The term *mini* relates to the narrow but detailed scope of
each specification.

A mini-specification accompanies every action symbol of the
data-flow diagrm. Each mini-specification is identified by the same
number as its corresponding action symbol. For example, the Corn-
Popper Data-Flow diagram of Figure 3.8 has an action symbol
labeled, "1.1 Collect Data." The mini-specification for this action
would be identified as, "1.1 Collect Data Mini-Spec."

The form of the mini-specification should be as simple and
concise as possible. It is the primary document used by software
engineers to create programs that accomplish the desired action. It
is usually written in a structured format called structured English
to conform to modern programming objectives of portability and
documentation. If structured English is used, the programmer has
an easier job converting the specification into the specific structured software language(s) used for the system as described by
Stiller (1978).

Structured English consists of a set of key works written in
a format that adds meaning to the key words. The most common
format is indentation. Every level of indentation indicates a subset
of commands that are to be executed until a logical conclusion is
reached. The basic key words and their logic are shown in Figure
3.12.

The level of command line indentation, combined with the
key words, conveys the logic form to be used. Other ordinary

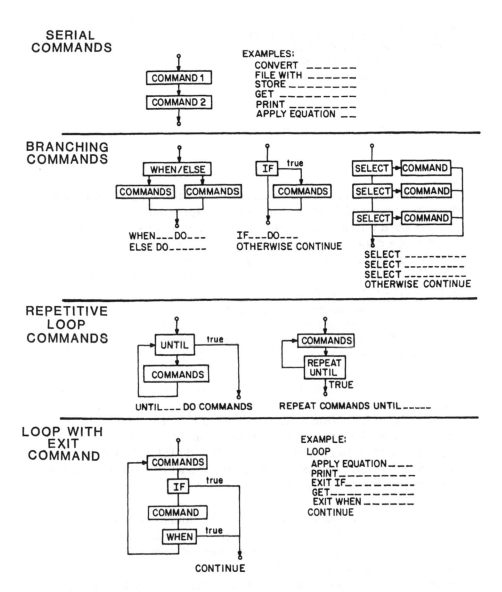

Figure 3.12   Structured English key words and logic.

Request temperature reading from popper
Store temperature in data record
Request date and time from clock
Store date and time in data record
Store data record on disk
Continue

Figure 3.13   1.1   Collect data mini-spec.

English words are also used to increase the information content of
the commands and to make the specification more intelligible.

There are four basic forms of logic required to define the
actions of a data-flow diagram. They are serial commands,
branching commands, repetitive looping commands, and looping
commands with internal exits. These various forms are discussed
in the following paragraphs.

Sequentially executed serial commands are written as a
column of left justified statements at the same level of indentation.
Figure 3.13 illustrates a mini-specification composed of the serial
command sequence for the action symbol "1.1 Collect Data" of
Figure 3.8.

Decisions are a major function of any computer-based manufacturing system. Decisions involve branching to the various alternatives available for action after a test is made. The decision logic
and key words for branching commands are also shown in Figure
3.12. The key word combinations of IF and OTHERWISE or
WHEN and ELSE describe single choice branches. Multiple choice
branches can be made by combining single branches or by using
the key word SELECT.

Branching commands are needed for many applications. For
example, suppose when the corn popper manufacturing system
tries to store the data record referred to in Figure 3.13, the power
to the disk is turned off and the store can not be executed. The
system designer should take care of this contingency by adding a
branch to the end of the mini-specification. The command
sequence shown in Figure 3.13 would then appear as Figure 3.14.

The third form of logic is repetitive looping. This is the repeated execution of the same command or series of commands

Request temperature reading from popper
Store temperature in data record
Request date and time from clock
Store date and time in data record
Store data record on disk
If data record is stored
    Continue
Otherwise
    Blow whistle for 10 sec.
    Shut down popper
Continue

Figure 3.14   1.1   Collect data mini-spec with branch.

until a test condition is met. The key words and logic for looping
are also shown in Figure 3.12. The test which ends the repetition
can be done before or after the series of commands is executed. If
it is done before, the key word is UNTIL or WHILE. If the test is
made after the loop is executed, the key words are REPEAT
UNTIL or REPEAT WHILE. As an example, to make the whistling
more effective the mini-specification could read the way it is given
in Figure 3.15.

Request temperature reading from popper
Store temperature in data record
Request date and time from clock
Store date and time in data record
Store data record on disk
If data record is stored
    Continue
Otherwise
Until 6 times
    Blow whistle for 10 sec.
    Wait 60 seconds
Shut down popper
Continue

Figure 3.15   1.1   Collect data mini-spec with branching and whistle control.

Request temperature reading from popper
Store temperature in data record
Request date and time from clock
Store date and time in data record
Store data record on disk
Loop
        If loop done 6 times
            Shut down popper
            Exit loop
        Exit if data record is stored
        Otherwise
            Blow whistle 10 sec.
            Wait 60 sec.
            Exit if operator shuts off popper
        Otherwise continue loop
Continue

Figure 3.16   1.1   Collect data mini-spec including branching, whistle
control, and shutdown.

The commands within the loop in this example are designated
by indentation under the "until"-condition test statement. After
the indented loop has been executed six times, the program exits
the loop and shuts down the popper.

The fourth form of logic needed for manufacturing system
control is the loop with exit shown in Figure 3.12. To illustrate
loop with exit, suppose a response is required from the popper
operator prior to shutting down the popper. This would be a way
to save the in-process product. Since the operator may on occasion
not be present when the whistle blows, it is also necessary that the
whistling be limited to six blasts and not continue indefinitely.
Finally, it would be desirable to shut down the popper automatically if the operator does not return to avoid damage to the equipment. The mini-specification could now appear as in Figure 3.16.

The mini-specification thus becomes a very concise document
of detailed commands configured in a format consistent with
modern structured programming. The format also imposes modularity. This is best illustrated by the logic diagrams of Figure 3.12.
Each of these diagrams have only one starting and one ending

point. Between the start and end points, a module has been defined which is independent of any other logic intrusion. This independence, like the modularity imposed by the data-flow diagram, makes software changes and growth much easier.

3.5 THE DATA DICTIONARY

The data dictionary describes the contents of files and documents represented on the data-flow diagram. These files and documents are usually defined on the diagram and in the data dictionary by the same letter-number combination. The letter identifies the type of data stored in the file or the type of document and the number identifies the specific contents. Common letter prefixes used are:

D = data from manufacturing operations, machines, and so on
S = standards for the manufacturing operation that are
    relatively constant
C = configuration data describing the contents of the factory
PR = process recipes
R = reports

The user defines these prefixes to meet the needs of the particular system. An example of a data dictionary is shown in Figure 3.17.

The alpha-numeric code to the right of dictionary entries defines the type and precision of the data stored. These codes are not the same as FORTRAN field specifications that include the spaces for signs, decimal points, and so on. The user can also define these codes to meet specific needs. The codes used in this example are:

A = use alpha-numeric character
I = use integer numbers
F = use fixed-point (decimal) numbers

The number associated with these letters tells how many characters or numbers will have to be stored for each entry. For example, A8 means the entry for this data is composed of 8 alpha-numeric characters. I2 means two integers.

For fixed point, the number to the left of the decimal tells how many whole integers are used and the number to the right of

C1   Production calendar
          For each year (AB)
              For each day (AB)
                  For each shift (I1)
                      Shift length in hours (F2.1)

C2   Configuration file
          For each machine in plant
              Machine ID (I2)

S1   Recipe to model cross index
          For each recipe ID (A4)
              Model name (A10)

D1   Machine data
          For each machine
              For each shift
                  Machine ID (I2)
                  Shift date (A8)
                  Shift ID (I1)
                  For each recipe used
                      Recipe ID (A4)
                      Production output (I-4)

R1   Monthly production output report
          For each month
              For each model
                  Model name (A10)
                  Units produced (I-4)
                  Units per machine hour (I4)

Figure 3.17   Example data dictionary

the decimal tells how many digits are to the right of the decimal.
For example, F2.3 means it is a fixed-point number with two
whole integers and three digits in the fractional part (For example,
54.358.) A minus sign placed after the code letter is used to show
that the number can be negative. The absence of a sign means only
positive numbers can occur. For example the Units Produced reported in R1 is shown as (I-4). This means production is anticipated
to be as large as tens of thousands requiring four integers to
quantify the magnitude. The (–) means that because of scrap, the
magnitude of the units produced can possibly be negative.

The data structure is inferred by the form of the data dictionary. Like the mini-spec, the data dictionary uses indentation to denote subsets of data. This relationship is a particularly valuable guide to setting up the owner/member relationships of the data base or file system used for data storage.

During the needs analysis phase of a Factory Information Systems design, the data dictionary grows along with the data-flow diagram, mock-ups, and mini-specs to describe the required system. The data dictionary becomes the document that describes the content of the manufacturing data base. It is used as the guide for data base design because it inherently contains not only the required data but the data structure (relationships).

3.6 THE MOCK-UP

The mock-up is a prototype report, chart, graph, or plot that describes by example the exact form and content of the documents to be produced by the system. The mock-up is very easy for system users to appreciate and understand. The popcorn production system produces a control chart, R1, for the superintendent (see Figure 3.8). The mock-up of this chart is shown in Figure 3.18.

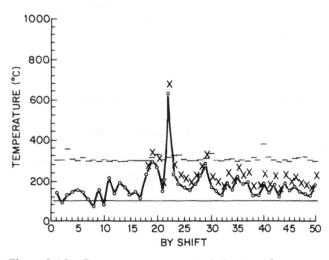

Figure 3.18   Popcorn popper control chart mock-up.

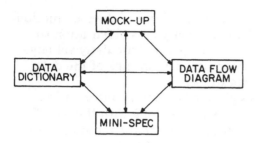

Figure 3.19   Procedure for structured systems design.

3.7 PROCEDURE FOR IMPLEMENTATION

Doing a structured system design using data-flow diagrams, mini-specifications, data dictionaries, and mock-ups is not a simple, straightforward, serial set of tasks. The value of this methodology is its recognition that the creative planning process occurs from looking at a problem both broadly and in detail, and from various viewpoints simultaneously. A technique the author uses to accomplish this is to lay out on a desk four piles of the above documents with work room in the center. Each pile contains the same type of document. The engineer achieves the system design by moving within the data-flow diagram and mini-specification piles to produce varying degrees of detail and from pile to pile for varying viewpoints. This technique is illustrated in Figure 3.19.

REFERENCES

DeMarco, Tom (1978). *Structured Analysis and System Specification*, Yourdon Press, New York.

DeMarco, Tom (1979). *Concise Notes on Software Engineering*, Yourdon Press, New York.

Stiller, Tom M. (1978). *FLECS: A Structured Programming Language for Minicomputers*, *RCA Engineer*, 23:(6), pp. 39-43.

4

Product Tracking and Flow Control

4.1 INTRODUCTION

This chapter describes techniques for following the movement of product (tracking) as it passes through the various manufacturing steps, techniques for controlling this movement (flow), and some of the ways of reporting product status. The diagramming of product flow which was not covered in Chapter 3 is also appropriately introduced here to aid in understanding its complexity.

The design of a FIS usually starts with the Product Flow diagram. These diagrams show locations where decisions have to be made regarding the disposition and future movement of product. The decision process, made at these locations, is defined and specified by applying the principles of data, information, and control flow discussed in Chapter 3.

Systems that do product tracking and flow control also supply information to many other financial, engineering, and manufacturing functions. This information is essential for these functions to be executed in a planned and controlled manner. The functions that depend on tracking and flow control can be appreciated by examining the basic manufacturing information flow diagram shown in Figure 4.1. First consider the upper left arm of this Y-like figure. It illustrates how the status of order completion is determined by tracking the movement of product related to these orders. That is why product tracking in a factory is a prerequisite to order tracking. To track this movement it is necessary to, (1) identify the product, and (2) know in a timely way where it is located in the factory. The tracking system collects data that contains product identification, location, status, and so on, and stores this information along with the time when it was collected in an appropriate data base. By relating the position of product in the process cycle to the orders that own this product, the order status is determined and projections about when the order will be completed can be made.

The right arm of Figure 4.1 supplies knowledge relating to the state of processing equipment. For example, is the machine operating or turned off, being repaired, running out of control, and so forth? This information is required, along with the knowledge of work location and the intended path of its movement, to control work flow. These two bodies of information come together at the juncture of the "Y" enabling humans or machines to control the flow of product.

Once product flow control is achieved individual manufacturing cells can determine what they are working on and can invoke the processing controls that are appropriate to the specific product being processed. Modern flexible manufacturing systems produce many different types of product in the same processing line. It is therefore essential that the product be identified.

To summarize, the control of manufacturing processes requires control of product flow. To achieve this the status of production machines and the location and type of the work must be known. If the location of the work is known, the status of orders can be determined. The functions of product tracking and flow control provide this capability. The systems that do this are

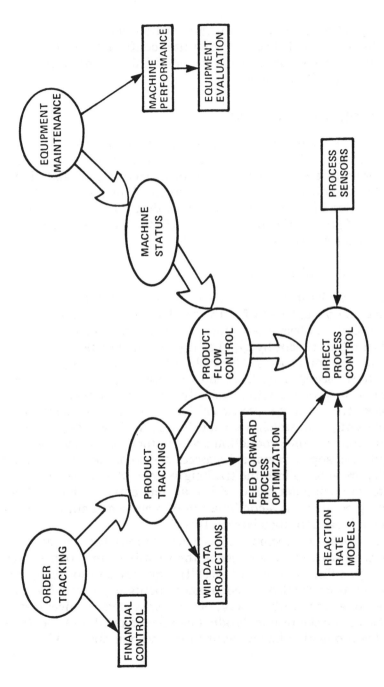

Figure 4.1 The Central Role of Product Tracking and Flow Control.

variously called tracking, monitoring, or flow control systems.
With the advent of distributed computer configurations for
factories these are more recently called distributed monitoring
and control (DMC) systems.

4.2 PRODUCT TRACKING

The task of tracking product requires a number of actions. These
are, (1) the identification of the product, including its relationship
to other entities such as orders, (2) the collection and validation of
data that describes the physical movement of the product, (3)
the collection of other information regarding the status of the
product, (4) the storing of these data, (5) the analysis, and (6)
the reporting of results.

4.2.1   Product Identification

There are two basic ways of accounting for product flow depending on how one identifies the product being manufactured.
The simplest, but least useful and accurate, counts product that
is unidentified (except by type) as it passes fixed reporting points
in the production process. This technique is sometimes called
"mass flow" accounting. A second way called "product tracking"
identifies the product with physically attached unique numbers
or codes. The product is physically identified as single pieces or
as groups of pieces which we will call lots. As these lots pass reporting points in the production process, their status (such as
quantities scrapped, reworked, passed, and so on) and their unique
identity are entered into the tracking system. Modern tracking
systems are computer-based with real time entry of data at the
reporting points using a combination of manual and automatic input techniques described later in this chapter.
      The mass flow technique has a very serious disadvantage.
There is no effective way to check the validity of the counts. The
inventory, mills, work-in-process (WIP), queues, and so forth, are
calculated from the difference in piece counts as product flows in
and out of an area with no way, other than a physical inventory,
to substantiate the results. Product can be scrapped, lost, or stolen
from the production line without timely accountability. Work

can also be moved to and from those infamous *informal* inventories that help make the operations books look like the accountants projections. This loss of accuracy is inherent with the absence of physical product identification. It does not reflect on the integrity of those reporting work movement. One innocent error, while entering data into the mass flow system, is propagated because there is no way to validate the quantity entered. Features for self correction, an absolute requirement in the design of a reliable computer system for data collection, are missing. The mass flow technique precludes incorporation of such features to force the correction of mistakes.

The accuracy of mass flow tracking is so poor that physical inventories must be taken evey month in high volume plants that produce many different product types. To accomplish this, the product must be held without movement until the counting is complete. In a large factory this usually means one day of lost production every time a physical inventory is made.

The Product Tracking technique provides, by physically identifying the product, a means for self correction of counts and absolute physical accountability each time product movement is recorded. This occurs because, when a lot is moved, the transaction software can be written to force the accounting to be balanced. The data must be correct and validated before the entry is accepted by the system.

The product tracking procedure is best described by an example. A small portion of a semiconductor wafer manufacturing operation was chosen as the example because it contains all the possible product moves experienced in other types of manufacturing.

At the beginning of a product move transaction, the identity of the lot of semiconductor wafers (pieces) is entered into the system. The tracking system knows the path that has been followed by this lot. It retrieves the lots' piece count from the entry made at the previous reporting point. It thus knows how many pieces started into the process that is currently being reported. All pieces of the lot must now be accounted for. If the lot had 100 pieces moved forward from the last transaction, 100 pieces have to be reported during the current transaction. If 80 pieces are to be moved to the next process step, the balance of 20 must be accounted for. For instance, 10 can be scrapped, 5 recycled, and 5 lost—but the sum of all categories reported must be equal to the number of pieces in the lot when it leaves the last reporting position.

The transaction validation software makes sure that the disposition of all pieces in the lot is reported. If the accounting does not balance, the system can warn, ask for a reentry, or, at minimum, log the discrepancy. (In actual practice the lost-and-found category becomes the way to report errors without slowing down the movement of product. Management by exception reports for lost-and-found provide rapid detection and solution of reporting problems.)

The people who are doing the wafer processing enter the movement data via video terminals. If invalid data is entered, a warning and request for correction of the input data is accomplished by blinking or highlighting the screen entries that could be incorrect. The potentially incorrect entries are determined by logical algorithms. The terminal will not accept the data until validation has occurred.

Experience with this type of product flow control has been very good. After six months of use, factory managers find it more accurate than physical inventories. They then eliminate these inventories and the accompanying loss of production time. Accountability is achieved because a means of reporting the realities of lost-and-found and the types of reprocessing, restoration, recycling, or repair is provided.

The system knows where product is physically located and the sequence of process steps the product has actually been through to get there. This is particularly helpful when the customers' product reliability requirements necessitate documented traceability as, for example, with the manufacture of heart pacers or missile components.

The cycle time required for each identified lot to travel through any part of the process is accurately determined This is done by computing the difference in the reporting time stamp that is stored with the transaction data. This cycle time is an important parameter to understand because WIP inventory is dependent on the sum of the components of cycle time which are the times spent in process, transit, and queue between process steps. In some types of manufacturing, like large-scale integrated circuits, process yield is also cycle time dependent. Because the measurement and control of cycle time has such a direct bearing on profit, product tracking is gaining greater acceptance. Before reliable computer systems were available to support this type of real time shop floor accounting, the product tracking technique was too slow and labor intensive. Now it is not only fast but economically justifiable.

4.2.2   Physical Identification Techniques

Product pieces or lots are identified in a number of ways depending on the manufacturing processes and the level of automation. In anticipation of later automation, if not to directly support it, almost all identification is chosen today to be machine readable. This means the identifier can be sensed and read or programmed by some device without requiring human intervention. These identifiers can take the form of barcodes, magnetic strips, small attachable magnetically programmable memory circuits, or simple mechanical appendages such as a coded series of holes or spikes. The choice depends on the processing environment and the product transport system. For close to room temperature environments information can be encoded in magnetic strips or solid state memories. Pressure sensitive barcode labels are available that can withstand soldering temperatures up to approximately 350°C. These are frequently used to identify printed circuit boards. Chemical etching, printed glass frits, or laser scribing is used to provide barcodes for products subject to even higher temperatures. Some examples of these products are vacuum tubes, semiconductors, hybrid circuits, and various devices made from glass or ceramic materials.

Barcode labels can be produced by computer driven printers located on the factory floor. This provides a way of selecting the information content of the barcode to match a particular current manufacturing need. Unfortunately, lower cost printers do not produce very good quality labels and do not overcoat the labels with protective coatings that are generally required to maintain readability through the manufacturing process. A better quality and ultimately cheaper method is to have the labels printed by an outside printing house. This is because the cost of good quality printing and overcoating equipment is significant. Barcode quality is essential for consistent first pass reads. Emphasis should be placed on attaining a high rate of first pass reads. This is necessary to prevent slowing down the flow of product and adding labor to the reporting operation.

There is no reason for a barcode, used to identify product, to contain any information. It only has to be a unique identifier. The computer system can easily relate many parameters to this unique identifier. Encoding the barcode with information like the product type is therefore a waste of space and an added complication. If specific information is encoded, large stocks of preprinted labels are required that have to be correctly selected during application. Simple, uniquely coded labels can be automatically applied to product by machine. For these reasons, it is better to use serially coded, high-quality labels and let the computer maintain the relationships to other parameters.

Correct reads are easy to achieve because the barcode is scanned many times during an interrogation and the result is only accepted if there is very good data consistency from each scan. The read will therefore either be correct or there will be no read. This happy circumstance makes barcodes very appropriate for product identification as long as quality labels and reliable readers are used.

Magnetic strips or small magnetically alterable memory circuits are identifiers that can be encoded under program control. They are sometimes attached to product and reprogrammed as the product moves through the process. This type of identifier permits the flexibility of having at least a small part of the product's data base travel along with the work as processing occurs. It can be used to identify the product and also to remember the last step in the process that the product passed through. Programmable identifiers are more fragile, however, requiring care in application, handling, and control of exposure to high temperature and the magnetic fields found around many types of processing equipment.

Identifier Location

In most manufacturing systems the problem of bringing the reader close to the identifier must be solved. The reader or scanner must be located reasonably close to the identifier being read for the period of time required to capture the encoded data. In a moving product transport system the read must be accomplished as the product passes the reader. Sometimes this is difficult to achieve, especially in flexible manufacturing systems, because the geometry of the different types of work varies, imposing constraints that prevent consistent identifier location. An example is the assembly of printed circuit boards where many different types of boards are intermixed in the production line. If these boards are of different sizes and shapes, it is impossible to standardize on a fixed position for the identifier. The readers or boards must then be moved relative to each other to achieve the read.

A solution to this location problem is to use two types of identification for the product. The first is attached to the product itself. The second is located at a fixed position on the carrier used to transport the product. Since the carriers always travel in a dimensionally controlled way, the fixed position of the carrier identifier can be scanned by readers located close to the transport system. Of course the identity of the product must be related to the identity of the transport carrier. This is usually done by manually reading the product identification with a wand when the product is placed in the carrier. The product and carrier identity are then in the system data base. From then on the product can be identified by reading the carrier identifier. The manual operation required to get the product and carrier identifiers entered into the computer system needs to be done only once and is valid as long as the product continues to move with the carrier. In most cases the product is carried through many process steps where tracking and control are required.

4.2.3   Product Flow Diagrams

The description of work movement is clarified with the use of product flow diagrams. The technique of product flow diagramming uses simple graphical terms to describe the paths of movement and the path options that have to be controlled. Various forms of product flow diagrams are used by the many commercially available software programs that model and simulate product flow. Each different simulation tool uses it own set of symbols. Some very elementary symbols are suggested here to illustrate the technique. It is strongly recommended that the FIS designer be familiar with the simulation of product flow and use this technique to uncover bottlenecks and other problems as part of the design process.

The product flow diagrams show the acceptable paths for product movement through the processes, tests, and decisions defined by the process specification or router. They do not contain information related to the flow of data, information, and control directives because the overlaying or mapping of these different flow diagrams onto each other results in a diagram that is much too complex to work with. The design problem is made tractable by dividing it into understandable components. The starting point is with the product flow diagram.

Two basic symbols are suggested for this diagram. They represent, (1) the steps of the process and, (2) where reporting into and out of the system occurs. These basic symbols are shown in Figure 4.2. Since reporting is usually not done after each process step, the steps that are not reported are grouped into a string of processes. The steps 1,2,... n in the box are executed contiguously without reporting results or making decisions regarding the disposition of the work or the direction of future work movement. The string can contain any number of process steps. Collecting processes into strings is used to simplify diagramming and in some cases to represent manufacturing cells.

The diamond symbols, called reporting points, are used to show the place in the flow of product where decisions are made and information is exchanged with the tracking system. The various actions that can take place at a reporting point are illustrated in Figure 4.2. The transactions at reporting points can be executed by processing and inspection machines or by people. In either case, there is a need for the validation of data entered into the system and the comprehension of data returned by the system to the sender.

A typical example of product flow is shown in Figure 4.3. This example covers a small part of a product flow diagram for integrated circuit manufacturing. In this case product is in the form of silicon wafers. Each wafer contains many integrated circuit "chips." The wafers are grouped together into lots containing a few wafers to a few hundred wafers. The first process step defined

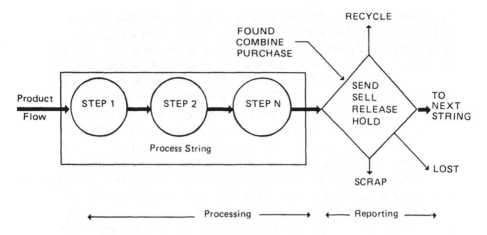

Figure 4.2   Basic product flow diagramming.

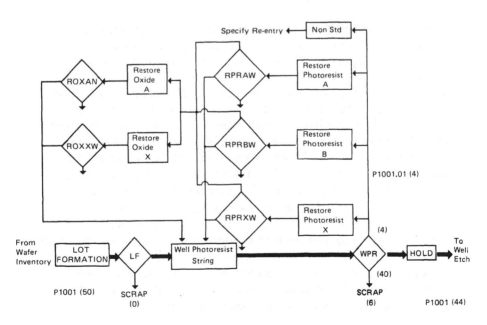

Figure 4.3   Integrated circuit manufacturing example.

by the product flow diagram is that of creating the lots from a wafer inventory. When the lot is formed, it is given a physical identification (P1001 in this example). The required information about the number, source, and history of the wafers that make up the lot is also entered. The entry of this information occurs at the reporting point shown as "LF." The handling of thin delicate silicon wafers occasionally results in some of the wafers getting broken. Thus the LF transaction must include the possibility of reporting some scrap. This contingency is shown in the product flow diagram by the arrow from LF pointing down to "scrap."

After the LF transaction is complete the lot moves on to the Well Photoresist (WPR) string of process steps shown in Figure 4.3. These steps are: application, bake, align, exposure, develop, bake, and inspect. When these steps are completed the results are reported at WPR and decisions are made regarding what to do with the wafers. They may move to a number of different restoration processes or on to the next base line process, Well Etch.

4.2.4   Reporting Movement

Once work is identified, the possibility exists for its movement to be followed. In this section we illustrate by example how this movement is reported. The example uses video data terminals as input/output devices and human operators as the source of data and the sinks of information produced by the system. These kinds of systems are often referred to as manual data entry systems. They should be avoided if possible because of the labor required. In many types of manufacturing this is not entirely possible because people are the only sources of some of the required information. Every condition and constraint that applies to a manual system also applies to automatic systems. The advantage of describing a manual system as the example, is the ease with which the reader can relate to the input and validation of data. Manual data entry systems also provide communication from the tracking system to the operator. If the communication contains control directives the system is also exercising control through the operator. The operator therefore must achieve both accurate data transmission and comprehension of data received.

ACTION _____ EMP _____ ENTRY _____ LOT _____

Figure 4.4A   Blank action line screen.

The initial display on the screen of the video terminal used to access the system is shown in Figure 4.4a. Four data fields are used: Action, Employee, Entry, and Lot. In this example the video terminals are programmed to work in a block mode. In this mode an entire screen of data is transferred to the computer system in one burst rather than transmitting character-by-character. This has the advantage of reducing the communication processing load on the computer system.

Initially the screen is blank except for the top line which has the above set of defined character fields. This top line is called the "action" line because it is used to specify for the system what action the operator wants to perform. The type of action has to be initially specified because the product tracking and control system is distributed throughout the plant and the same terminals are used to request many different types of information as well as enter it from these many locations.

The operator first enters the action to be performed which, in the example of Figure 4.3, is lot formation. LF is therefore entered in the action field as shown in Figure 4.4b. It is desirable to know the identity of the person doing the reporting. This is because experience has shown that the person entering the data should be the one who has done the work. The reason becomes clear when data validation is considered. Invalid data, detected by consistency or logical checks during entry validation, can only be corrected by those who did the work and realistically know what happened.

The reporting person enters his or her identity as employee number 61616. No other data has to be entered for the system to

ACTION LF_____ EMP 61616__ ENTRY _____ LOT _____

Figure 4.4B   Action line screen filled in.

ACTION <u>LF</u>　　　EMP <u>61616</u>　ENTRY ＿＿＿＿＿＿ LOT ＿＿＿＿＿
　　　　LOT NUMBER <u>P1001</u>＿

　　　　TYPE NUMBER <u>TCC193</u>＿＿

　　　　OXIDATION LOT # <u>X650</u>＿　＿＿＿＿＿　＿＿＿＿＿

　　　　WAFERS REMOVED　<u>50</u>

　　　　WAFERS　PASSED <u>50</u> LOST ＿＿＿　FOUND ＿＿＿

　　　　WAFERS SCRAPPED FOR:　OTHER　WARPED　DIMPLED　BROKEN
　　　　　　　　　　　　　　　　＿＿＿　＿＿＿＿　＿＿＿＿　＿＿＿＿

　　　　REMARKS <u>EXAMPLE LOT</u>＿＿＿＿＿＿＿＿＿＿＿＿＿＿＿＿＿＿＿

Figure 4.5　Lot formation screen.

know what to request of the operator to complete the LF transaction. The operator therefore sends the action line screen to the
computer by pushing one of the terminal keys.

　　A program determines the action requested and based on that
knowledge selects a second screen which asks for the rest of the
data required to complete the formation of the lot. The new screen
is written by the computer on the face of the terminal monitor
below the action line so both "windows" are simultaneously viewed
by the operator as shown in Figure 4.5.

　　In our example the computer system is programmed to supply
the lot number. This ensures that it is unique. When the screen of
Figure 4.5 appears it presents the assigned lot number, P1001, to
the operator. This permits the operator to write this number on
any paperwork or traveler that goes along with the lot. If barcodes
or similar identification devices were used, the operator would
attach the barcode label generated by an adjacent printer or would
wand in the number of a preprinted label and attach it to the work.

　　Lot formation is used to relate the lot to a particular design
or type of product. This is done by entering the type number,
TCC 193. If desired the lot could also be related to a specific
order or market requirement.

Next, the source of the wafers is entered as X650 along with the quantity of wafers removed from inventory. The number of wafers passed is 50, which is equal to the number of wafers removed from inventory. If some of the wafers had been broken as the lot was physically assembled, the difference between wafers passed to the next string and wafers removed from inventory would have to be accounted for as scrap. The reasons for scrapping wafers is also entered. As one might guess from the screen, the reasons for rejection can be "other," "warped," "dimpled," or "broken." The last entry is called "remarks." The remarks entry permits operators to enter information about problems that had not been anticipated by the system designers. These remarks can be retrieved by engineering and management to help increase their awareness of system or process problems. As solutions are developed they can be incorporated in the manufacturing system. This provides a means for constant system improvement through evolution.

When the LF screen information is received by the computer system, it determines the next process step by referring to the process router or recipe for the "type" previously entered. The screen shown in Figure 4.6 is then displayed to the operator. The system is thus controlling the flow of product by directing it to follow a previously defined process path. Multiple process path options are illustrated in the next example reporting point, Well Photoresist. Well Photoresist (WPR) is a series of processes used to apply a photosensitive pattern on each wafer. This pattern is used as a mask against chemical etching in subsequent processes. The application of a patterned photoresist is done by the string of steps shown in Figure 4.3. These steps are the cleaning of the wafer surface, application of photoresist material, a bake to remove solvents, the alignment of an optical mask with the existing pattern on the wafer, exposure through the mask of the photoresist layer with ultraviolet light, development of the exposed pattern using solvents, a bake to harden the remaining photoresist, and inspec-

NEXT ENTRY FOR "PASSED WAFERS"...WPR

Figure 4.6   Control directive for product flow.

tion of the final pattern. After this series of processes is completed, the wafer lot moves to the Well Photoresist reporting point, WPR.

During the string of Well Photoresist processes the wafers may be processed without mishap, or some problem may cause some of the wafers to be defective. They then have to be reworked or scrapped. Depending on the nature of these problems various actions need to be taken. The Well Photoresist reporting point thus requires analysis of inspection results, decision-making based on these analyzes, and the generation of control directives for what should next happen to the wafers. This analysis-decision-control generation cycle occurs at almost every reporting point. The computer system, with appropriate programming, assists with the consistent execution of this cycle. Artificial intelligence is now being applied to make these analyses and decision making tasks easier and more consistent.

Figure 4.3 shows that some wafers from lot P1001 could be seriously damaged so the option exists for them to be scrapped. Some could be passed on to the next set of base-line process steps. Some of the wafers may suffer damage to the photoresist layer which is not injurious to the wafers themselves. They can therefore be restored to their original state by removing the photoresist, inspecting the results and, if all is well, returned to the Well Photoresist string of processes. In our example a sublot, P1001.01, has been formed at WPR to accomplish this. It is also possible that after the photoresist is removed, it is found that there has been damage to the underlying oxide layer. There is then the option of restoring this oxide by removing and replacing it before reentering the Well Photoresist process string. Each of the restoration processes is followed by a reporting point where, as in the case of the base line process, the results of prior processing are reported and the future path of product flow is determined. All of these possible product flow paths are shown in the product flow diagram of Figure 4.3.

Sometimes it is desirable to restore the damaged wafers from a lot and hold the balance of the wafers until the restoration process is completed. The restored wafers are then combined with the original lot. This is done if the homogeneity of the lot is unimpaired. The objective is to reduce the number of small lots that must be handled by the manufacturing system. In our example the undamaged wafers identified as lot P1001 are held at the reporting point until the restored wafers identified as lot P1001.01 have finished reprocessing and are reported again at WPR. When this occurs the screen that returns to the operator after the WPR transaction for lot P1001.01 directs that lot P1001 be released, combined with lot P1001.01, and the combination moved as P1001 to the next process. Lot P1001.01 then ceases to exist. However lot P1001.01 remains uniquely defined in the data base so that its history can be retrieved.

If wafers cannot be combined after a restoration because it is felt that the wafer damage will impact the uniformity of wafers in the lot, and therefore the integrity of any analysis, the sublot which was generated and used for identification of the wafers that are restored is maintained. In this case, the acceptable portion of lot P1001 continues to move forward as P1001. The sublot P1001.01, which contains the restored wafers, will move as a separate entity proceeding independently through its processes.

The entry of data at WPR is accomplished by using the video data terminal screen shown in Figure 4.7. The action here is to

ACTION SEND   EMP 61616   ENTRY WPR   LOT P1001

    WAFERS RECEIVED   50

    WAFERS PASSED   40

    WAFERS   LOST _____ FOUND _____

    WAFERS FOR: MAL PDF MRP WRM BDM TPR INC FMR OTH WPD DPD BKN

       SCRAPPED   ___ ___ ___ ___ ___ ___ ___ ___ ___ 4 2 ___

       RECYCLED   3 1 ___ ___ ___ ___ ___ ___ ___ ___

    RECYCLED TO: RESTORE RPRA ___ or RPRB X OR RPRX ___

             OR NONSTANDARD _____ REENTER AT _____

    REMARKS THIS IS A TEST TO SEE IF YOU ARE AWAKE

Figure 4.7   Well photoresist data entry screen.

"send" lot P1001 from entry WPR to the next step or steps of the
process. In this example we will show how the accounting of
wafers is achieved and restoration is monitored. Note that the
action is always directing the movement of product from a reporting point. This provides the system with the knowledge of
where the product will next be reported. In the case of multiple
path options the appropriate ones are specified as will become apparent. This knowledge of where product is going is essential for
the accounting of lot quantity and for the control of product
flow.

The 50 wafers that were sent from Lot Formation to Well
Photoresist are entered at the WPR reporting point as "wafers
received." If less or more had been received the difference would
have to be entered as lost or found by the operator or the system
would not accept the transaction. In our case, 40 wafers passed
inspection and could go on to the next process step. The difference
between the 50 wafers received and the 40 wafers that passed
inspection must be accounted. In Figure 4.7 four of these wafers
were scrapped because they were warped. The operator enters a
"4" in the field of the screen marked "WPD." (The screens are
programmed to prompt the operators for all known dispositions,
reasons for failure, and so on, associated with the particular entry
or reporting point. These prompts appear when the screen is first
generated before the operator is asked to enter any data.) Two
other wafers were also scrapped because they were dimpled
"DPD." The remaining four wafers were recycled, three for misalignment "MAL" and one for a photoresist defect "PDF." The
sum of the scrapped and recycled wafers now equals the difference
between the wafers received and the wafers that passed inspection.

The last entry on this screen is the control directive for the
restoration process to be used in removing the photoresist from
the recycled wafers. Since some wafers were allocated for recycling, the system requires that one or more restoration processes
be entered and that these are acceptable for the stated reasons for
recycling. The system thus imposes a discipline of acceptability on
the designated restoration procedures. For misalignment and
photoresist defects the restoration process designated by "RPRB",
Restore Photoresist using Restoration process B, is satisfactory.
When the operator enters this to define the flow path of the four

HOLD "PASSED WAFERS"
ASSIGN "RECYCLED WAFERS" LOT # P1001.01
NEXT ENTRY FOR "RECYCLED WAFERS" ... RPRB

Figure 4.8   Control directives for lot disposition.

recycled wafers, it is therefore accepted. If an unacceptable restoration process had been specified, the entry would blink after the
screen of data was sent to the computer, indicating to the operator
that the restoration process RPRB entry could not be validated.
The system will not accept a screen of data unless all of the required entries are made and all portions of the data are consistent
and validated against whatever logic can be developed from existing
knowledge.

    After the "WPR" screen is accepted as valid the return screen
would appear as shown in Figure 4.8. In this case the decision (by
a rather intelligent system) is to hold the passed wafers because
the system knows that wafers failing for misalignment and defects
are not seriously damaged. The passed wafers can thus be combined
with the restored wafers without loss of lot "integrity." The lot
identification for the recycled lot (P1001.01) is shown so the
operator can physically mark the lot being recycled. The specific
acceptable process for restoration is shown as RPRB to designate
the restoration path for those wafers. The system will expect four
wafers to arrive at the RPRB reporting point and force the
accountability of these wafers.

    After the wafers are restored by process B they are again in-
spected and the results entered at the reporting point, RPRB. This
data entry screen is shown in Figure 4.9. The restoration process
was successful so the four wafers can be sent from RPRB to the
base-line process. To show operator 55555 that this is the next
flow path for the product, the return screen of Figure 4.10 after
the "send" from RPRB specifies the next process step, WPR. It
does this by accessing the knowledge base or recipe file that contains the only acceptable alternative for the results reported.

    After WPR processing the wafers are passed through the
"WPR" reporting point where they previously had failed. This is

ACTION <u>SEND</u>   EMP <u>55555</u>   ENTRY <u>RPRB</u>   LOT <u>P1001.01</u>

   WAFERS RECEIVED <u>4</u>

   WAFERS PASSED <u>4</u>

   WAFERS LOST _____   FOUND _____

   WAFERS FOR   UXD VOX PAR OTH WPD DPD BKN
      SCRAPPED   ____ ____ ____ ____ ____ ____ ____

      RECYCLED   ____ ____ ____ ____

  RECYCLED TO: RESTORE ROXA _____
               RESTORE ROXX _____
               NON STD. _____ REENTER AT _____

Figure 4.9 Disposition after RPRB restoration.

NEXT ENTRY FOR "RECYCLED WAFERS" ...WPR

Figure 4.10 Control directive showing re-entry point to base line process.

shown in Figure 4.11. The return screen of Figure 4.12 then informs the operator that the wafers from this lot P1001.01 can be combined with the wafers from lot P1001 which are being held at WPR and that the combined lot, again designated as P1001, should move to the Well Etch process, "WETCH."

The reporting point transaction for a product tracking system must therefore be capable of accepting a number of different actions. These actions are: sending forward to the next base-line process, losing or finding components of a lot, scrapping portions of a lot, recycling portions of a lot to alternative but valid restoration processes, selling a lot out of the system and purchasing a lot from another system, holding and releasing a lot, combining a lot or portion of a lot with another lot or lots, and splitting a lot into sublots. The software to do this has to handle these transactions, maintain accurate accountability of the lot constituents, and validate each data entry.

ACTION <u>SEND</u>   EMP <u>61616</u>   ENTRY <u>WPR</u>   LOT <u>P1001.01</u>

    WAFERS RECEIVED <u>4</u>

    WAFERS PASSED <u>4</u>

    WAFERS LOST _____ FOUND _____

    WAFERS FOR MAL PDF MRP WRM BOM TPR INC FMR OTH WPD DPD BKN

      SCRAPPED ____ ____ ____ ____ ____ ____ _____ ____ ____ ____ ____

      RECYCLED ____ ____ ____ ____ ____ ____ ____ ____ ____ ____ ____ ____

    RECYCLED TO: RESTORE RPRA_____

                  RESTORE RPRB _____

                  RESTORE RPRX _____

                  NON STD. _____REENTER AT _____

Figure 4.11   Pass of restored wafers.

COMBINE WAFERS FROM LOTS

P1001

P1001.01

    AND

ASSIGN THEM LOT #. . .P1001

NEXT ENTRY FOR "PASSED WAFERS" . . .WETCH

Figure 4.12   Disposition of combined lot.

Some typical input/output screens for video terminals used to do these transactions have been illustrated by Figures 4.4 to 4.12. There are other possibilities for lot movement, process selection, accountability, and so on. The important concept to convey is that each manufacturing domain has rules, some better defined than others, that allow humans to decide what to do next in various situations. Computer systems, using conventional as well as techniques based on artificial intelligence, can remember and invoke these rules to help discipline the control of very complex manufacturing operations.

4.3   PRODUCT FLOW CONTROL

The preceding section on product tracking showed how product flow directives were generated as the result of transactions that take place at the reporting points. These directives are necessary and sufficient for controlling the movement of product as it leaves these reporting points. There are often, however, many process steps between the reporting points. These steps can be grouped as strings for the purpose of tracking but it is necessary to control the flow of product within these strings without directives from the product tracking system.

Most strings can be defined as work cells. A work cell contains all the process steps that are required to make a logically defined or measurable change in the product. It is, therefore, reasonable that reporting be done after these changes occur. The processes within a cell are usually very tightly coupled so that their control is very dependent on what is going on within the cell and only weakly dependent on the status of things outside the cell. The following sections examine the control of product flow into a cell from a reporting point and control within the cell between reporting points.

4.3.1   Movement Following a Reporting Point

By definition a product reporting point is required whenever the flow of product can change. This includes any of the decisions other than the send forward shown in Figure 4.2. Transport systems must have access to these control directives to know what product should be delivered to which location next. When manual transport systems like carts or fork lift trucks are used the operators of these vehicles get direction from the tracking system as illustrated in Figures 4.6, 4.8, 4.10, and 4.12.

Automated transport systems like conveyors and automatically guided vehicles (AGV) must also have access to these same directives. In addition they must know the route to follow and, in the case of AGV's, which delivery to make next. The product flow control system supplies this information by making these decisions with the help of some reasonably complex algorithms.

Product location on conveyor systems is determined by using machine readable identifiers attached to the product or to a related carrier. These can be read as the product enters a junction where its path may have to change. The tracking system determines ahead of this junction what the new path should be and activates the mechanism for changing direction or removal. Wired communication is used between the readers, activating mechanisms, and the tracking system.

Communication to automated guided vehicles is usually done by microwave or infrared transmission systems. The AGV often maintains its own data base related to what it contains. The flow control system can then interrogate these vehicles, analyze the next moves, and direct the vehicles to accomplish these diverse tasks.

The flow control system is therefore an integral part of transportation control between tracking system reporting points. Product movement within the processes between reporting points has to be controlled by some other means.

4.3.2   Movement Between Reporting Points

Between the product tracking reporting points we are in the realm of the manufacturing cells. They are controlling their processes and they must control the flow of work within these processes. These are highly sophisticated subsystems that are very domain dependent. It is therefore difficult to generalize about how they should be designed because the design is highly dependent on the physics of what is being done. There are some generic reasons for accomplishing flow control within the cells which are worth reviewing.

The cost saving and improved market responsiveness achieved by reducing work-in-process (WIP) is well documented. The reduction of WIP forces a change from the mass production of a few product types to the mixed manufacturing of many different products made on the same production line. Manufacturing systems designed do this are called flexible manufacturing systems (FMS).

They will use people and machines very inefficiently unless they are precisely scheduled and dynamically controlled. The real time product tracking and flow control systems discussed in this chapter are, therefore, essential for effective FMS production.

The product flow control system for a FMS is designed to cope with four needs: (1) respond to a change in schedule, (2) avoid inoperable equipment and the destruction of product or machines, (3) optimize machine utilization, and (4) assure safety and prompt machine maintenance.

Respond To Schedule Change

Dynamic real time scheduling is altering the way factories are managed by forcing the re-routing of product during manufacturing. These requirement-driven scheduling systems referred to in Hadavi (1985) need the kind of responsiveness that can be achieved only by reaching down to the cell level of the factory. This is happening as illustrated in the following printed circuit board assembly example.

In this example of flexible manufacturing, printed circuits boards (PCB) are assembled using automatic component insertion machines. These machines contain or can get on demand all the recipes for every possible type of PCB board they will assemble. Every time a new board type enters the machine the appropriate recipe is used for controlling the selection and placement of inserted components. The machine knows what recipe to use by reading a barcode label attached to each board which defines the board type.

A change in schedule or a shift in priority simply requires the insertion of the new boards into the component insertion machine. The machine learns it must change its process for inserting components and determines what the new recipe is when the new barcode is read. Once machine readable unique board identity is established, smart algorithms can be applied to control the sequencing of boards entering a machine to minimize the setup time and/or to keep a series of closely coupled insertion machines optimally utilized.

Avoid Inoperable Equipment And Destruction of Product or Machines

Sensors in equipment inform the cell controller that the equipment is not functioning or that it is operating out of control. Product flow control within the cell then diverts the product to an alternative machine or prevents it from entering the machine where damage to the product or machine can occur.

Wrong, mis-conveyed, or missing product can damage machines. This can be avoided if detected and the introduction of this product is stopped. The action of many machines used for FMS is automatically initiated by the introduction of work pieces. If other supporting functions for the machines such as oil pressure, power to subsystems, die temperature, or missing parts are detected it is wise to stop the entrance of product to avoid damage.

Optimize Machine Utilization

The ability to manage the flow of work in cells containing a number of flexible machines will determine how well the physical resource and the people that run it are utilized. This is a scheduling problem that has such short lead time that it borders on control. That is why product flow within the cell has to be done in real time by the cell control computer.

A good example of machine utilization control is the selection of various types of mixed printed circuit boards to go into a cell of component insertion machines. When this facility runs flexibly with very small lot size, the working time of the machines can be improved if the incoming boards are selected based on the time it takes different boards to be processed in each machine. The selection algorithm endeavors to keep some boards in the queue of each machine so no machine stops running for lack of product. The same algorithm also tries to limit the queue size ahead of each machine.

Safety and Maintenance

Material movement can be dangerous to both operators and machines. Many industrial accidents involve material rather than the machines that process it. An effective flow control system should include features that automatically stop the movement of material when humans are present in locations that pose a threat to safety. The classic example of this is the material transport robot that is protected from humans by light beams, trip wires, or enclosures with sensed openings.

The product flow control system within a cell can also notify maintenance personnel when material is not flowing into a machine. For whatever reason, this may give them time to accomplish short term corrective or scheduled maintenance tasks and warn them that an unscheduled shutdown has occurred.

REFERENCE

Hadavi, K., (1985). *Dynamic Scheduling for FMS*, Conference Proceedings of Autofact (1985). Computer and Automated Systems Association of the Society of Manufacturing Engineers, Dearborn, Michigan, p. 6-59.

5

Statistical Control Charting

David Coleman

*The author wishes to acknowledge and
express his thanks to David Coleman
for providing this chapter.*

5.1 INTRODUCTION

One of the greatest problems facing factory managers is determining what and where problems are occurring as production takes place. Most factories are too large to locate problems personally and in a timely fashion. Installing a computer system that collects millions of bits of data does not help either, because there is not enough time to look through the stacks of printed material produced by such a system and manually analyze the contents. Fortunately there is a solution to this problem. It is the concept of management-by-exception: analyze the data by computer and report only those situations that are classified by the analysis as problems. There are two useful techniques for this analysis. One is the use of statistics to search the data for significant deviation from normal. The other is the use of artificial intelligence to detect problems that are less well defined and more heuristic in nature. This chapter explains some of the statistical methods used for the non-intuitive data driven initiation of management-by-exception reports by computer systems that contain these statistical analysis tools. The following chapter will introduce the reader to artificial intelligence.

5.2 CONCEPT OF STATISTICAL CONTROL CHARTING

5.2.1 Plotting Data Over Time

In concept, statistical control charting involves plotting historical data from a process versus time (Figure 5.1)—to watch the change over time of a measurement of a process or performance characteristic or a rate, such as proportion (or percent) defective. To this plot are added control limits (sometimes called action limits) which are statistically determined, and which help us to decide whether the process is stable, or is unstable and should be adjusted (Figure 5.2).

For example, process A may output steel ball bearings which are nominally 1 cm in diameter. Ordinary process variability which affects process A (and any other process) makes it impossible to produce bearings which are all exactly 1 cm in diameter. Some amount of variability is a fact of any manufacturing process. That is why there are +/- specifications for manufactured products. Though a process may be deliberately and carefully designed to produce items with dimensions or other characteristics at some target value, there will always be some distribution about the target in every measurement of quality that is used. This is true even when measurement error is disregarded. Specifications reflect engineers' informed judgment of how much variability about target values can be tolerated without serious impact to product acceptability or performance.

Suppose process A produces one ball bearing every one hundredth of a second. If we sample every 200th ball bearing from production, and measure its diameter (ignoring measurement system error and variability in diameter due to bearing orientation), we can build up a diameter-measurement chronology of our process. What if we were to plot each measurement against time; one point for every two seconds of production? This time-plot, or

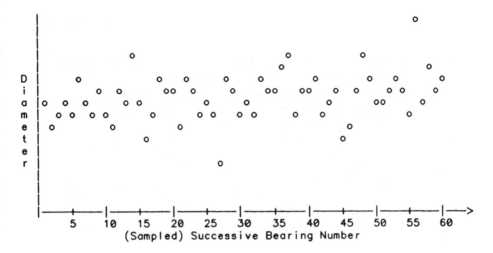

Figure 5.1   Time plot of samples successive bearing diameters.

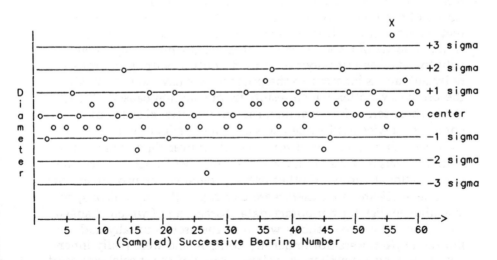

Figure 5.2   Control chart of samples successive bearing diameters.

time series, shown in Figure 5.1, would reveal some of the variability in the production process. The data points do not all lie on the horizontal line which corresponds to the target diameter, 1 cm.

Added to this plot, a statistical control chart (Figure 5.2) has a set of horizontal lines, control limits, which are used to indicate whether or not the process is in control. These lines are statistically determined, based on certain assumptions made about the distribution of the data. A set of rules is used with the control limits to judge the most recent sequence of plotted points at any given time, and to declare the process In Control if no rule is triggered, and Out Of Control if one or more of the rules is satisfied.

5.2.2 Normal and Abnormal Behavior

What does it mean for a process to be in good statistical control? It means that the process is behaving well, process measurements are staying at about the same level, well within what we would expect, and the proportion of product rejected is what we expect it to be. More specifically, the process is not affected by special causes of product variation, only common causes. A special cause of variation is a major contributor to variation over which we have some control, and which is not inherent to the process. Special causes of variation should be eliminated or at least have their effect on product variation reduced. A common cause of variation is the opposite of a special cause. It is usually not a major contributor, is often not under our direct control (without a process redesign), and is inherent to the process. We have to put up with the product variation which results from common causes, at least in the short term. For example, the computational speed of a computer service used while I was an undergraduate was affected by several things; it showed some variation over time. Some of the common causes of variability were the number of other users, the computational burden of other users' programs, the proportion of sharable software other users were using, and the amount of space available on disk for temporary data storage, all of which changed over time. These were things which in themselves usually had limited impact, were not under my control, and were really inherent to running a timesharing system. Some of the special causes of variability were when someone once dropped an ice cream cone into one of the computer cabinets, which crashed the system, and when someone got the computer tied up in an infinite loop, which caused the Central Processing Unit to become preoccupied with chasing its tail uselessly for some while.

Some causes of variability are hard to classify. If the choice of steel alloy for the ball bearings affected bearing diameter in process A, then we would classify steel alloy as a cause of variation. But whether it would be a special or a common cause would depend on the standard operating procedures followed in the process. If many alloys were used, and they were all similar in how they affected diameters, then we would call steel alloy a common cause. Suppose, however we used alloy 1 three quarters of the time and alloy 2 one quarter of the time, and suppose alloy 2 caused a change in diameter that was most of the tolerance range (difference between specification values). Then we would call steel alloy a special cause of variation. We would be inclined to stop using alloy 2 because of the effect on product quality.

We say that a process is in statistical control if it is not inflicted by special causes of variation, only common causes. However, because there are grey areas in classifying causes of variation, there are grey areas in calling a process in or out of control. Most processes are neither grossly out of control nor in tight control, but somewhere in-between. A process tightly under control typically has time-plots which are at a roughly constant level, and without the following:

Trends in the overall level

Sudden shifts in the overall level

Periodicity in the level (that is, cycling up and down)

Changes in variability (that is, changes in the range of values of the data)

Any or all of these patterns are typically present in the data from a process which is grossly out of control. To a lesser degree, they are seen in a process which is only loosely under control.

An absence of such patterns reflects consistent behavior over any different but similar sources and circumstances in the manufacturing process from which the data came, such as: different shifts, lines, machines, testers, workstations, workers, operators, setups, suppliers, vendors, and so on. These are things which change over time, but we do not want them to cause our product quality to change over time.

5.2.3 Quantifying Behavior—Control Limits

How do we decide whether a process is in control, using a control chart? Any process which is grossly out of control will show it-self to be out of control in a simple time-plot. Often, though, production processes go out of control in a manner not obvious to the user of the time-plot. Using control limits, statisticians have developed ways to objectively decide whether current data indicates a statistically significant out of control situation — to which someone should respond. The simplest case is a measure-ment which is too large or too small to be explained under normal process conditions. Statistically, we can compute what too large and what too small are, and any time a measurement is beyond those extreme values, we can with certainty blame it on a special cause, a monkey wrench in the system. This judgment can be coupled with particular circumstantial or engineering knowledge to help decide what should be done. For example, we may know that a fixture tends to lock up, or a test lead tends to make poor contact, or that a rapid change in environmental conditions had occurred at a certain time.

5.2.4 The Relationship of Control Limits to Specifications

A common misconception, one which drives statisticians crazy, is that control limits and specifications are the same thing, and should be used the same way. This is false. Control limits are not equivalent to specifications. Extreme control limits reflect what a process should be able to do — after all, that is by defi-nition how they are calculated, as we will see below.

Specifications reflect what we want a process to do, whether or not it can. We hope to be working with processes whose ex-treme control limits are well within the specifications. Such a process is said to be capable of making product to specification, or simply, a capable process. Processes whose extreme control limits fall near the specification lines are marginally capable, and need to be brought under tighter control by eliminating or reducing special causes of variation. Processes whose extreme control limits are far beyond the specifications are incapable, and need to be brought under much tighter control, or (more likely) need to be substantially redesigned to eliminate or reduce causes of variation which are common to that process.

5.3 STATISTICAL DISTRIBUTIONS

5.3.1 Histograms and the Likelihood Curve

Most people are familiar with histograms but are not sure how to interpret or use them. Histograms are important to the understanding and use of statistical control charts because they help us to better understand the distribution of measurement data from a product or process. Suppose we had the following data, measurements of transformer inductance on a group of transformers, produced by process B:

22.6977	19.7303	16.2276	21.1880	19.6443	22.7566
19.2593	20.7488	16.8322	17.8873	18.3072	27.2195
23.8054	19.9496	18.2283	19.6950	21.5448	22.9028
19.9410	18.5408				

These twenty measurements can be displayed in a histogram, as in Figure 5.3.

Though we cannot get a detailed picture of transformer inductances from this process, using this histogram, we can safely say that the process produces transformers of inductance about 20, with most in-between 16 and 24, and probably in a single-peaked (unimodal) distribution about 20. For comparison, we note that this seems a bit different than another process, process C, which gives a histogram of 20 transformer inductances which looks like Figure 5.4.

Notice that while each histogram is centered on the same value of about 20, and has most of the measurements occurring between about the same values, the shapes of the histograms are a bit different. This might imply a difference in process. In fact, perhaps process B is better controlled than process C; process C has greater variability. We might guess that there were two differ-

```
        PROCESS B
MIDDLE OF    NUMBER OF
INTERVAL     OBSERVATIONS
    16         1      *
    17         1      *
    18         3      ***
    19         2      **
    20         5      *****
    21         2      **
    22         1      *
    23         3      ***
    24         1      *
    25         0
    26         0
    27         1      *
```

Figure 5.3 Histogram of twenty measured transformer inductances from process C.

ent process factors affecting process C. Perhaps process C has two different types of magnet wire, or is produced during two different shifts, or involves the use of one of two fixtures at a time, and so on.

5.3.2 Getting Close to the True Parent Distribution

By taking more measurements, we can get a clearer idea of the true parent distribution of transformer inductances, as approximated by histograms. The parent distribution is the exact, underlying distribution. It is the distribution we would ideally obtain

```
        PROCESS C
MIDDLE OF    NUMBER OF
INTERVAL     OBSERVATIONS
    16         3      ***
    17         3      ***
    18         2      **
    19         1      *
    20         0
    21         0
    22         1      *
    23         6      ******
    24         4      ****
```

Figure 5.4 Histogram of twenty measured transformer inductances from process C.

by collecting an infinite number of measurements from the process, under normal operation. If we increase our sample size to 100, we get Figure 5.5 for the transformer process B and Figure 5.6 for the process C.

The common bell shape of the Gaussian (normal) distribution seems to characterize well the inductance measurements of magnets produced by process B. A rather different, bimodal shape is seen in measurements from process C. When the sample size is increased to 1000, the histograms in Figures 5.7 and 5.8 approximate the underlying distributions rather well, and we can even spot a little asymmetry in the magnitude of the peaks in process C.

5.3.3   Models and Approximations to the Parent Distribution

As occurs in any science, much is gained in statistics by developing models which approximate the real world. The histograms we can construct from process data could—given enough time, effort, and stability of process—come arbitrarily close to approxi-

```
         PROCESS B
MIDDLE OF     NUMBER OF
INTERVAL      OBSERVATIONS
    11          0
    12          1      *
    13          1      *
    14          0
    15          3      * * *
    16          8      * * * * * * * *
    17          7      * * * * * * *
    18         10      * * * * * * * * * *
    19         16      * * * * * * * * * * * * * * * *
    20         15      * * * * * * * * * * * * * * *
    21         11      * * * * * * * * * * *
    22          8      * * * * * * * *
    23          8      * * * * * * * *
    24          5      * * * * *
    25          0
    26          4      * * * *
    27          2      * *
    28          1      *
```

Figure 5.5 Histogram of 100 measured transformer inductances from process B.

```
            PROCESS C
MIDDLE OF    NUMBER OF
INTERVAL    OBSERVATIONS
    14          1        *
    15          1        *
    16         11        ***********
    17         20        ***********************
    18         10        **********
    19          1        *
    20          1        *
    21          4        ****
    22         16        ****************
    23         15        ***************
    24         14        **************
    25          5        *****
    26          1        *
```

Figure 5.6   Histogram of 100 measured transformer inductance from process C.

```
            PROCESS B
EACH * REPRESENTS    5 OBSERVATIONS

MIDDLE OF    NUMBER OF
INTERVAL    OBSERVATIONS
    10          1        *
    11        . 0
    12          0
    13          8        **
    14         15        ***
    15         25        *****
    16         49        **********
    17         86        ******************
    18         82        *****************
    19        141        *****************************
    20        154        ******************************
    21        125        *************************
    22        106        *********************
    23         86        ******************
    24         49        **********
    25         34        *******
    26         20        ****
    27          6        **
    28          8        **
    29          4        *
    30          0
    31          0
    32          0
    33          0
    34          1        *
```

Figure 5.7   Histogram of 1000 measured transformer inductances from process B.

112

```
          PROCESS C
EACH * REPRESENTS    5 OBSERVATIONS

MIDDLE OF    NUMBER OF
INTERVAL     OBSERVATIONS
   14           1     *
   15          31     *******
   16         107     *********************
   17         178     **********************************
   18         103     ********************
   19          28     ******
   20           9     **
   21          32     *******
   22         139     ***************************
   23         206     *****************************************
   24         129     *************************
   25          34     *******
   26           3     *
```

Figure 5.8 Histogram of 1000 measured transformer inductances from process C.

mating the true, parent distribution of process measurements. Usually, though, we do not have enough time, cannot justify the effort, and do not have stable enough processes to accurately characterize each process measurement distribution uniquely.

Instead, we rely on models, in the simplest sense of the word model. We compare histograms of process data to known statistical distributions, and if they are close enough in shape we assume that the parent distribution for that measurement data follow that distribution: that distribution becomes our model for the parent distribution. Alternatively (as discussed below), we can use sample means of successive groups of N measurements, which have a distribution approaching the Gaussian as N goes to infinity. This alternate strategy has the Gaussian distribution as the model of the true distribution of group means. A model can never exactly characterize a process, so accuracy suffers. However, we gain the ability to generalize to many measurements, from many distributions. If we know the properties of the statistical distributions we use for models, then we have a body of knowledge on which we can draw to make a statistical decision as to whether or not the process is in control. Note that we are making a statistical decision when we assume that a process measurement follows a certain distribution. We cannot make the assumption simply because we do not have any alternative strategies.

5.3.4   Statistical Distributions that We Work With

The Binomial Distribution

The binomial distribution, more than most others, has particular mechanistic or physical appeal to the engineer and scientist. This is because the underlying assumptions on which it can be based are simple and are frequently believable. They are: there is a device or process with only two outcomes, where

> —outcome (a) has probability p of occurring with each use (or fire, or cycle) of the device. Since there are only two outcomes, this imples that
>
> —outcome (b) has probability q = 1 − p of occurring with each use;
>
> —the kth outcome does not in any way depend on the (k − 1st) outcome, or on any previous outcome; that is, the outcomes are independent;
>
> —moreover, p is constant over time.

If these assumptions hold, then the binomial distribution can be used to describe, evaluate, and predict the behavior of the device. They hold, or nearly hold, for much of the machinery used in manufacturing. Using the binomial distribution, the probability of getting k out of n fires to result in outcome (a) is,

$p(n,k) = C(n,k) \times (p^k) \times (1 - p)^{(n-k)}$, where C(n,k) is

> "n choose k,"
>
> $= n!/[k!(n - k)!]$

(Note: j! is "j factorial" = $1 \times 2 \times \ldots \times j$.)

For example, rolling a fair die to get a particular number is appropriately modeled by the binomial. The chance of getting 4 rolls of a "1" out of 10 rolls of a die is

$$p(4) = C(10,4) \times ([1/6]^4) \times ([1 - 1/6]^6)$$

$$= 1 \times 2 \times \ldots \times 10/[(1 \times 2 \times 3 \times 4) \times (1 \times 2 \times 3 \times 4 \times 5 \times 6)]$$
$$\times ([1/6]^4) \times ([1-1/6]^6)$$

$$= 210 \times .00077160 \times .334898$$

$$= .054265$$

The power of the binomial distribution in control charts (particularly p-charts) is that we can assess the likelihood of observing certain binomial events, such as sequences of pass/fail, accept/reject, or go/no-go. For example, if a wealth of carefully collected historical data indicated that a jelly donut machine produces jellyless donuts only $p = 1/1000$ of the time, then the likelihood that we would find three jellyless donuts out of the last five donuts produced, given the above binomial assumptions, is

$$p(5,3) = C(5,3) \times (.0001^3) \times (.9999^2)$$

$$= 1 \times 2 \times 3 \times 4 \times 5/[(1 \times 2 \times 3) \times (1 \times 2)] \times (.0001^3)$$
$$\times (.9999^2)$$

$$= 10 \times (10^{-12}) \times .9998$$

$$= 10^{-11}, \text{ approximately}$$

In this situation, without even going into all the mathematics of why the binomial distribution takes its particular form, or how to compute $C(n,k)$, we could say with great certainty that one of the binomial assumptions does not hold. The assumption we would suspect to have been violated is the "constant p" assumption, which states that the probability of getting a jellyless donut is always .0001. We have strong evidence that this probability has increased during the last five donut machine cycles—for some reason. Perhaps we are out of jelly. The binomial distribution is used with p-charts, as described below.

The Gaussian Distribution

The Gaussian (unfortunately, also named normal) distribution is well known primarily because it is ubiquitous. It seems to appear (or at least people seem to want it to appear) almost everywhere that measurements are taken. Tables of the standard Gaussian distribution appear in every elementary statistics textbook. See Figure 5.9 for a plot of the Gaussian. It is commonly assumed

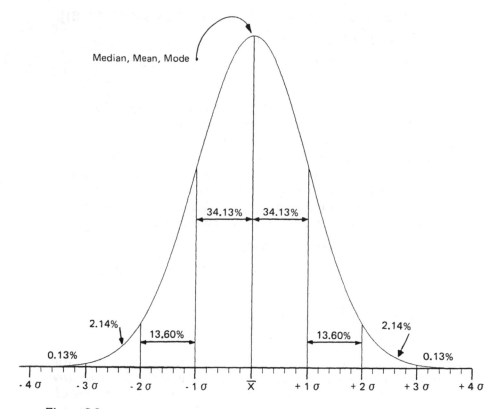

Median, Mean, Mode

34.13% 34.13%

2.14%

13.60%   13.60%

0.13%   0.13%

2.14%

- 4 σ   - 3 σ   - 2 σ   - 1 σ   X̄   + 1 σ   + 2 σ   + 3 σ   + 4 σ

Figure 5.9

to be a good representation (model) for measurement data because of a very weak (and actually theoretically incorrect) application of the central limit theorem (CLT). The correct CLT states:

The sum of N independent random variables which are identically distributed with finite mean and finite variance > 0, and finite third moment, is distributed ever-closer to a Gaussian distribution (converges in distribution) as N increases.

This theorem is commonly assumed to hold for measurement systems where natural variation is observed and is (probably appropriately) attributed to many roughly independent sources, each with small contribution, but where the effect or distribution of each source's contribution is unknown. In these situations, the finiteness assumptions are usually not violated, but the sources of small contributions to variability do not behave similarly, so their distributions are not identical. Thus, the CLT cannot be correctly applied. Even so, roughly bell-shaped distributions of measurements *are* in fact common.

Whether or not measurement data from a process are distributed according to the Gaussian distribution, the CLT *can* be used for sample means (group means) of successive groups of independent sequential measurements. Note that the sample means of successive groups are not the same as moving averages. Moving averages are serially correlated, hence, not independent. As an example, suppose measurement data for fiber tensile strength were distributed as in Figure 5.10, for a type of carbon fiber.

These data are unimodal, but are definitely skewed to lesser values. However, histograms in Figures 5.11 and 5.12 of means of successive groups taken from this set of data show a shape which draws ever closer—as the sample size increases—to the bell curve. The Gaussian distribution and the CLT are used with X-bar charts, as described below.

The Poisson Distribution

The Poisson distribution is famous for modeling defect counts on units of fixed dimension, where the number of opportunities or potential sites of a defect is great for any unit, but the likelihood of a defect at any given site is very small. Of course, rare events other than the occurrence of a defect can be modeled. Examples of areas of application of the Poisson distribution are:

1. Ink blots in a newspaper

2. Microscopic cracks in a car's windshield

3. Loops dropped in a knitted sweater

4. Mold cultures on a piece of bread

```
      CARBON
EACH * REPRESENTS   5 OBSERVATIONS
MIDDLE OF      NUMBER OF
INTERVAL       OBSERVATIONS
       0          1     *
       5         31     *******
      10         88     *****************
      15        121     ************************
      20        121     ************************
      25        143     ****************************
      30         95     ********************
      35         70     **************
      40         70     **************
      45         50     **********
      50         35     *******
      55         29     ******
      60         34     *******
      65         28     ******
      70         17     ****
      75         11     ***
      80         15     ***
      85          3     *
      90          7     **
      95          8     **
     100          6     **
     105          1     *
     110          1     *
     115          2     *
     120          3     *
     125          4     *
     130          1     *
     135          1     *
     140          0
     145          1     *
     150          0
     155          0
     160          0
     165          0
     170          0
     175          0
     180          1     *
     185          1     *
     190          0
     195          0
     200          0
     205          0
     210          0
     215          0
     220          0
```

Figure 5.10 Histogram of carbon tensile strength.

```
          CARBON5
MIDDLE OF    NUMBER OF
INTERVAL     OBSERVATIONS
    15           2      **
    20          23      ***********************
    25          39      ***************************************
    30          38      **************************************
    35          29      *****************************
    40          26      **************************
    45          17      *****************
    50          14      **************
    55           5      *****
    60           5      *****
    65           1      *
    70           0
    75           1      *
```

Figure 5.11   Histogram of average carbon fiber tensile strength—groups of 5.

```
          CARBON10
MIDDLE OF    NUMBER OF
INTERVAL     OBSERVATIONS
    20           3      ***
    25          16      ****************
    30          32      ********************************
    35          18      ******************
    40          15      ***************
    45           9      *********
    50           4      ****
    55           3      ***
```

Figure 5.12   Histogram of average carbon fiber tensile strength—groups of 10.

5.   Knots in a length of used kite string

6.   Dandelions in a front yard

7.   Typos in a manuscript

The most famous example of the use of the Poisson is Bortkewitsch, who used the Poisson distribution to model the number of Prussian soldiers who were killed by the kick of a horse in each of 10 cavalry corps in each of 20 years. The results were:

Deaths from Kicking of a Horse	Actual Frequency	Poisson Expected Frequency	Discrepancy Between Actual and Expected
0	109	108	1
1	65	66	− 1
2	22	20	2
3	3	4	− 1
4	1	1	0

The Poisson distribution is used with p-charts, as described below.

5.4 PROVEN AND POWERFUL CONTROL CHART METHODS

5.4.1 X-bar Charts, for Measurement Data

An X-bar chart is a time-plot of arithmetic means (averages) of
successive groups of measurement data, plus control limits added
to help determine whether the underlying process is in control.
The reasons for plotting successive means rather than the raw
data in time order are: (a) the data are condensed, (b) more per-
sistent (not fleeting) trends are identified, and most importantly,
(c) the distribution of the successive means is usually better understood than the distribution of the raw data (the underlying parent
distribution is not usually known). This latter point is an application of the well-known statistical theorem already discussed, the
central limit theorem (CLT). We now discuss this theorem in
more detail, then apply it.

To review, the CLT states that, "The sum of N independent
random variables which are identically distributed with finite
mean and finite variance > 0, and finite third moment, is distributed ever closer to a Gaussian distribution (converges in distribution) as N increases." The assumptions that must be made
for the theorem to be applicable are almost always valid. A finite
mean is the case for any data encountered in the real world.
Finite, nonzero variance is almost always the case; textbook distributions, such as the Cauchy, violate this assumption, but are
rare. Finite third moment refers to a bounded measure of skew-
ness, which is always satisifed in the real world. The CLT says
that the distribution of the sum converges to a Gaussian. This

also means that the distribution of sample means converges to a Gaussian (as N increases), and the mean of that distribution is the mean of the underlying parent distribution. Not only does the distribution of the sample means conform ever closer in shape to a Gaussian as N increases, but the width of the distribution shrinks inversely proportional to the square root of N (due to another theorem in statistics). The way we usually measure width is by the standard deviation, which is the square root of the variance. The variance is (1/N) × (the sum of the deviations of each observation from the sample mean)—squared. Hence, the distribution of sample means where N = 100 is closer to the Gaussian, and has a standard deviation one half that of the distribution of sample means where N = 25.

Even peculiar far-from-Gaussian parent distributions have distributions of successive means that come closer and closer to a Gaussian distribution as the sample size, N, increases. The skewed carbon data distribution shown in Figure 5.9 were seen to have sample mean distributions which quickly become Gaussian in appearance as N increases. By using sample means and the CLT instead of raw data, we are intentionally throwing away information about the parent distribution. This is the price we pay for the ability to understand the behavior (distribution) of the sample mean—so we can judge which process measurements are unusual, and which are typical. This works because if the raw measurements increase (conversely, decrease) in value, a corresponding increase (conversely, decrease) will be seen in the mean of those measurements. The CLT says that sample means of an in-control process should be approximately distributed according to the Gaussian distribution, and a large enough shift up or down indicates that the underlying distribution has changed. The process is out of control; measurements are significantly trending or shifting from their historical level.

A simple example of the use of the CLT is the roll of a fair die. In this example, we actually know the parent distribution. In the real world, we do *not* know the underlying distribution, but this example is chosen so we can see how well the CLT actually performs. The chance of any one of the 6 outcomes of the roll of a fair die is uniform: 1/6 for each outcome. The mean outcome is the sum over all outcomes of: (each outcome) times (the probability of each outcome) = (1) × (1/6) + (2) × (1/6) + ... +

(6) × (1/6) = (1/6) × (1 + 2 + 3 + 4 + 5 + 6) = 7/2. The standard deviation is the square root of the sum over all outcomes of: [(each outcome − mean outcome)^2] times (the probability of each outcome) = [(1 − 7/2)^2 × (1/6)] + ... + [(6 − 7/2)^2 × (1/6)] = $\sqrt{35/12}$.

We compute these values only because they are of interest and useful in applying the CLT. No single outcome of the roll of a fair die could be labeled unusual, since each outcome has likelihood 1/6. If we observed a string of k 6s in a row (k > 1), however, we might suspect that the die was unfair, with suspicion rising with increasing k. The chance of observing k 6s in a row is (1/6) k. Alternatively, we could apply the CLT. Since in this example we have the luxury of knowing the parent distribution exactly, we can exactly compute the distribution of the mean of two successive rolls of the die, which is identical to the roll of a pair of identical dice, since dice have no memory. As any veteran Monopoly player knows, this distribution is a simple triangular: the probability of getting a mean of 1 or 6 is pr(mean is 1) = pr(mean is 60) = pr (sum is 2) = pr(sum is 12) = 1/36; similarly, pr(mean is 3/2) = pr (mean is 11/2) = 2/36, pr(mean is 2) = pr(mean is 5) = 3/36, pr(mean is 5/2) = pr(mean is 9/2) = 4/36, pr(mean is 3) = pr mean is 4) = 5/36, pr(mean is 7/2) = 6/36. Figure 5.13 shows how this is approximated by a Gaussian distribution with a mean of 5/2, and standard deviation of $\sqrt{35/12} \div \sqrt{2}$. Though the Gaussian distribution does not have a shape similar to the triangular distribution, the CLT has in no way failed in this case; the sample size, N = 2, is just too small to have effectively washed out the very non-Gaussian shape of the parent uniform distribution. Let us consider a larger sample size, say, N = 10.

For sample size N = 10 we can again compute the sample mean distribution exactly for the example, since we know the parent distribution exactly: the probability of getting a sample mean of (10 + i)/10 is $C(10,i)/(6^{10})$, where as above, C(10,1) = 10!/[i! × (10 − i)!] . Using the CLT with mean 7/2 and standard deviation $\sqrt{35/12} \div \sqrt{10}$, we get quite a reasonable approximation to the exact distribution, as can be seen in Figure 5.14. This approximation is good enough so that we can exploit the fact that G(0,1), the standard Gaussian distribution with mean = 0 and variance = 1, is well known and widely tabulated. We can do this

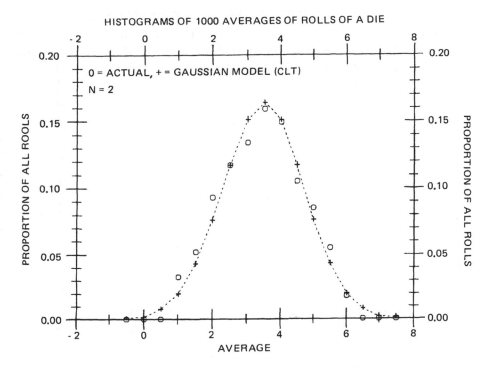

Figure 5.13   Histograms of 1000 averages of rolls of two dice.

because if random variable X (a sample mean) has a Gaussian distribution with mean mu and standard deviation s, then Z = (X − mu)/s has a standard Gaussian distribution. In our 10-rolls-of-a-die example, we let X = the mean of 10 rolls, then $(X - 7/2)/(\sqrt{35/12} \times \sqrt{10})$ is approximately distributed as G(0,1). An unusually long series of high rolls or low rolls will result in a sample mean X that is high or low relative to the approximating Gaussian distribution. A standard X-bar chart of the die would have one or more of its rules triggered, and the die would be declared out of control.

What constitutes an unusual value for the Gaussian distribution? Common sense and the shape of the Gaussian dictate that a sample mean far from the grand mean (population mean) is unusual. The grand mean is taken to be the historical average of in-

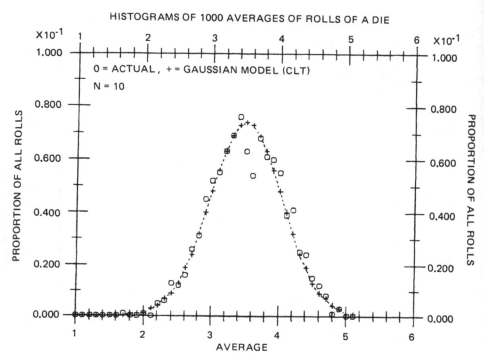

Figure 5.14 Histograms of 1000 averages of rolls of ten dice.

control data, and is plotted on the X-bar chart with the label mu, or X-double-bar, or center-line. As Figure 5.9 shows, the chance that a value drawn from the Gaussian distribution will be further than 3 standard deviations from the grand mean, on either side is 0.13 percent + 0.13 percent = 0.26 percent. This is the basis for what is called, Rule 1 in standard X-bar control chart parlance. A single sample mean further than 3 standard deviations from the grand mean triggers an out of control signal. This is often indicated by a large X mark on the control chart, placed above the offending point, as seen in Figure 5.2. Common sense tells us that other patterns should also be cause for concern—perhaps a sequence of points which contained no extreme points, but was unusually high or unusually low as a group. Rule 2 is 2-out-of-3 successive points more than 2 standard deviations from the grand mean. The chance of that happening under normal, in-control operating conditions is: [the number of distribution tails] times [the number of ways one can select one point out of two] times [the probability of two Gaussian-distributed points being further than 2 standard deviations from the grand mean on any given side] = 2 × 3 × (.0227) = .00153. Rule 2 is illustrated in Figure 5.15.

Rule 3 is 4-out-of-5-successive-points further than one standard deviation from the grand mean (chance = .002668, Figure 5.16). Rule 4 is 8 successive points in a row on one side of the grand mean (chance = 2 * (1/2) ↑ 8 = .0039, Figure 5.17). Another rule, which we will call Rule 5, is oriented toward the very long term: 17-out-of-20-successive-points on one side of the grand mean (chance = .001087, Figure 5.18). How to interpret a wide variety of patterns seen in X-bar control charts is discussed in Sec-

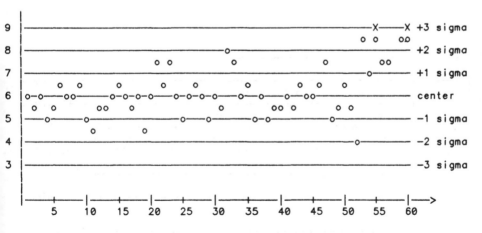

Figure 5.15 Rule 2: Two out of three successive points beyond 2 sigma from the mean.

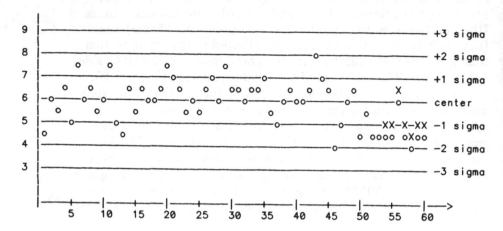

Figure 5.16 Rule 3: Four out of five successive point beyond sigma from the mean.

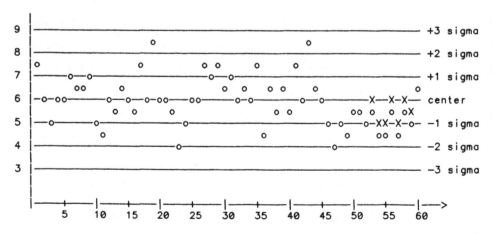

Figure 5.17 Rule 4: Eight successive points on one side of the mean.

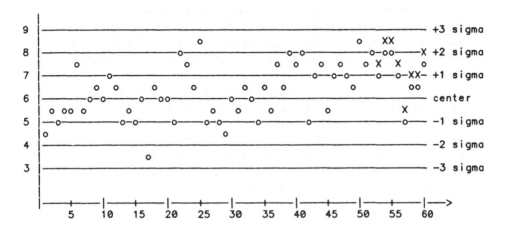

Figure 5.18   Rule 5: Seventeen out of twenty successive points one one side of the mean.

tion 4.3. The types of patterns shown will activate different rules or combinations of rules.

Strengths and Weaknesses

The X-bar control chart has several strengths. First, it is good for most process and measurement data. Since the CLT has distributional assumptions which are easily met, X-bar control charts can be widely used. Almost any parent distribution will work, as long as N, the size of the group for computing sample means, is large enough. Second, as with any good control chart, the X-bar chart provides both graphical display and quantitative analysis. Graphical display means that patterns over time will often be readily discerned. Quantitative analysis means that people will not react solely on hunches or subjective judgments, but will be able to motivate action on the basis that they are highly confident that a process is out of control (better than 99% confident, using Rule 1 alone). Third, the most common patterns of deviation from control, shifts and trends, are detected by X-bar control charts.

There are also some weaknesses to the X-bar control chart. First, it usually does not detect periodicity, such as daily cycles. Second, it is only an approximation—in two ways. First, the averaging process and application of the CLT give only an approximately Gaussian distribution of sample means. The only remedy for this problem is to use a larger sample size, or to explicitly model the parent distribution. Second, the Rules are (conservative) discrete approximations to the continuous Gaussian distribution. Statisticians can help remedy the second approximation by going to finer resolution than steps of a single standard deviation. The author is doing research in this vein on a more general approach, "Generalized Control Charting."

5.4.2 R-charts, for Measurement Data

An R-chart is a time-plot of ranges (maximum minus minimum) of successive groups of measurements, plus control limits to help determine whether the process yielding in the data is in control. The R-chart is a companion to the X-bar chart, and should be consulted *before* the X-bar chart. Abnormally high or abnormally low measurements may not be due to a shift or trend in level, but rather due to abnormally large fluctuations. The R-chart will detect such fluctuations.

The range, R, is always positive. Uncharacteristically large values are what concern us. They can signal that the process has gone out of control in the way we usually think of a process going out of control: outgoing product varies too much. The other extreme, unusually small values of R mean that the measurements are unusually consistent — which does not indicate an out-of-control situation that needs to be addressed. Therefore, only upper control limits are used (above the center or R-bar line). The same rules as are used in X-bar control charts can be used, but in many applications only Rule 1 is used. In R-charts, the control limits represent the same level of (un)likelihood as they do in X-bar charts, but they are not actually the R-bar value plus one, two, and three standard deviations. They are derived either theoretically or empirically as multiples of R-bar and do not rely on the CLT or the Gaussian distribution in any way. Multiplying factors to be used to compute the control limits of a variety of sample sizes are given in most books on statistical quality control.

The R-chart, or some suitable substitute, should be used in conjunction with an X-bar chart (or its suitable substitute) for several reasons. First, as mentioned, abnormal fluctuations may explain the presence of abnormally high and/or low values. Second, anyone who manages a process is most definitely interested in an increase in fluctuations of process or product characteristics *as well as* a shift or trend in average level. Such behavior is a genuine indication of being out of control. Consider the process of driving a car. If the steering system is biased to one side, you will want to have the car repaired—to center it again (X-bar). If the steering becomes unstable, with uncertain response when you turn the wheel, you will also want to have the car repaired—to tighten up the steering system (R). Third, as with X-bar charts, we can choose rules (with the help of a statistician) which will provide arbitrarily high confidence when declaring out of control (we should note that every time we add or modify a rule to increase our level of confidence when we declare out of control we pay a price: we decrease our level of confidence that the process is actually in control when the control chart says that it is). Fourth, out of control in R almost always means that more product is out of specification. Fifth, though a shift or trend in X-bar can often be corrected by changing a control variable setting or restoring the process to normal operating conditions, an increase in variability (larger R) often means a fundamental breakdown of the process.

Strengths and Weaknesses

The R-chart has several general strengths and weaknesses. First, it works with most process measurement data. Second, it provides both graphical display and quantitative analysis. On the negative side, it may not detect periodicity. Also, R-charts can be hypersensitive to uninteresting changes in the parent distribution. For example, negative serial correlation could be present in the process; a large value is most likely to be followed by a small value, and vice versa. This may be of no consequence, but it could easily trigger the R-chart rules. Less sensitive, robust alternatives to the R-chart have been and will continue to be developed.

5.4.3 P-charts, for Counted Data

A p-chart is a time-plot of values of p (usually rates of defective product), which are computed from successive groups of binary data (pass/fail, accept/reject, or go/no-go), plus control limits added to help determine whether the process resulting in the binary data is in control with respect to the p level. Basically, a p-chart is used to answer the question, "Is the level of defectives (or defects) that we are currently experiencing simply due to random variation, or is it significantly higher than normal?" An example used frequently by the author is the simple but surprising case of a part-making process running 10% defective (p = 0.1). Suppose parts are produced in lots of 10. What is the minimum number of parts in a lot of 10 which would signal (with high certainty) that the production process is out of control, and thus should be fixed before more parts are produced? I asked this question of people at all levels from operator to middle-level manager in a manufacturing plant. I had a custom-made sampling device to simulate the production of parts at the 10% defective level. How would the reader answer the question? Is 4 out of 10 high enough to indicate out of control? Should we demand more, say, 7 out of 10? The answer is given exactly by the binomial distribution, used in p-charting to compute the control limits. This requires that we believe the binomial distribution assumptions, as discussed in Section 2.4.1. They usually are valid, so let us compute in Table 5.1 the likelihood of seeing k defectives out of 10 parts, given a historical defective rate of 10%.

We can see that the likelihood of observing 4 or more defectives out of a group of 10 is 1 − .9984 = .0016 which is very small, under the binomial assumptions. Were we to actually get a lot of 10 parts with 4 defectives, we would suspect that the constant p binomial assumption had been violated. We would assume that p had actually increased, causing a greater number of defectives than can be statistically explained away as ordinary random variation.

P-charts can be used in a manner similar to X-bar charts, except that the lower control limits need not be used. An unusually low p value indicates an unusually small number of defectives, which is a favorable situation, not likely to require any control response. Typically, the control limits used correspond to the center plus one, plus two, and plus three standard deviations, and the traditional rules to determine in or out of control are applied. The limits are not necessarily determined by computing standard deviations, however.

Table 5.1

k	Probability of seeing k/10 defectives	Result	Cumulative
0	$(1 - 1/10)^{10} = .9^{10}$	.3487	.3487
1	$C(10,1) \times (1/10) \times (1 - 1/10)^9 = .9^9$	.3874	.7361
2	$C(10,2) \times (1/10)^2 \times (1 - 1/10)^8 = 45 \times [(.1)^2] \times .9^8$	.1937	.9298
3	$C(10,3) \times (1/10)^3 \times (1 - 1/10)^7 = 120 \times [(.1)^3] \times .9^7$	.0574	.9872
4	$C(10,4) \times (1/10)^4 \times (1 - 1/10)^6 = 210 \times [(.1)^4] \times .9^6$	.0112	.9984
5	$C(10,5) \times (1/10)^5 \times (1 - 1/10)^5 = 252 \times [(.1)^5] \times .9^5$	.0015	.9999
6	$C(10,6) \times (1/10)^6 \times (1 - 1/10)^4 = 210 \times [(.1)^6] \times .9^4$	.0001	1.0000
7	$C(10,7) \times (1/10)^7 \times (1 - 1/10)^3 = 120 \times [(.1)^7] \times .9^3$	(0)	1.0000
8	$C(10,8) \times (1/10)^8 \times (1 - 1/10)^2 = 45 \times [(.1)^8] \times .9^2$	(0)	1.0000
9	$C(10,9) \times (1/10)^9 \times (1 - 1/10) = 10 \times [(.1)^9] \times .9$	(0)	1.0000
10	$C(10,10) \times (1/10)^{10}$	(0)	1.0000

In a few limiting cases in statistics, certain distributions can be substituted as approximations for other distributions. This is true, in particular, for p-charts. The binomial distribution is the exact distribution for p-charts, and is the preferred distribution to use. More commonly, the binomial approximation to the standard deviation is used (the standard deviation of the mean of binomial data = $\sqrt{p(1-p)/N}$, then the Gaussian distribution is used to compute the control limits, as in an X-bar chart. This approximation is appropriate only when p is close to 1/2 (a rare situation), or when N is very large relative to p (Np should also be at least 5). Another approximation choice is the Poisson distribution, which is appropriate when p is small, N is very large, and Np is roughly constant. Like the binomial, the Poisson is a discrete distribution, and the distribution probability levels are a function of the number, k, of defectives rather than the proportion of defectives, k/N, so you have to convert k to proportion of defectives by dividing k by N.

The binomial and Poisson distributions do not necessarily give probability levels at the same percent points as those associated with the Gaussian mean plus one, plus two and plus three standard deviations. Therefore, to use the traditional rules of determining in and out of control, mathematical interpolation can be used to estimate the control limit values. Linear interpolation usually suffices: accumulate a partial sum of probability values, as in Table 5.1, then find the two probability values which bracket a desired probability value, and do the linear interpolation. The Gaussian control limit at the mean plus one standard deviation is at 84.13% = 0.8413, so we would interpolate between k = 1 and k = 2 in Table 5.1, since 0.8413 is between the corresponding cumulative probability levels 0.7361 and 0.9298. This is only worth doing if we are going to use this interpolated value as an approximation for samples of size close to N, rather than recomputing for each N. Otherwise, since it is impossible to get a fractional number of defectives, we use the larger value, k = 2.

Whatever distribution we use to compute control limits, some general guidelines apply. First, we want to use large samples (large N). If N is small, the control limits will be wide, more approximate, and more variable with N. On the other hand, the smaller the sample size, the more timely will be the control chart. Second, we want the sample sizes to be as nearly constant as possible. If N varies greatly, then the limits will noticeably vary, which makes interpretation of the control chart more difficult (a rule of thumb is to use an average N throughout if the variation in N is less than +/– 10% about the average N).

Strengths and Weaknesses

The p-chart has a few strengths, and more weaknesses. On the positive side, it can be used for most binary data. It provides both graphical display and quantitative analysis. It can detect shifts and trends. On the negative side, it usually will not detect periodicity. It will usually not detect changes in variability, and has no companion chart to do so—unlike the X-bar chart, which can be supplemented by the R-chart. It requires many data. Roughly speaking, a thousand or more binary values is equivalent to one process or product measurement value. Lastly, the p-chart can mask sophisticated or subtle underlying behavior.

5.5 PRACTICAL IMPLEMENTATION AND USE OF STATISTICAL CONTROL CHARTS

5.5.1 First Principles: Closing a Control Loop

The effective use of control charts comes about by institutionalizing control loops which contain them. A common pitfall is to not make this crucial step. The author has been to many plants where control charts were produced in a regular and disciplined manner, and taped to the walls or tacked to the bulletin boards, but never used. Control charts, by themselves, cannot solve problems. They need to be interpreted by people whose job it is (at least in part) to interpret them, and then the information they convey needs to be *acted on*. This requires that the interpreter (analyst) have the authority to take appropriate actions, for example, shutting down a line. The factory control discussed in Chapter 2 and the control and data flow diagrams discussed in Chapter 3 can effectively integrate the FIS with the control charts, the control chart analysts, and the action. This is called closing the control loop.

5.5.2 Process Capability Studies

Before instituting control charts, process data which is being considered for control charting should be well understood. This can be accomplished through a process capability study. The additional benefit of such a study can be the better understanding of the process.

A detailed description of a process capability study is beyond the scope of this book. In brief, a process capability study for continuous measurement data is accomplished by taking the following steps: (1) Collect data from a process over an uninterrupted interval of time—one which is long enough to include important possible sources of variations in the data, and sample frequently enough to detect transient phenomena. (2) While collecting the data record all possibly relevant process events, such as line stoppages, machine failures, changes in operators, materials, and so on. (3) Plot the data in time order and label all events. (4) Temporarily discard data associated with events that are extraordinary, special causes of variability. (5) If the data do not exhibit any peculiar behavior (such as illustrated in the X-bar chart Patterns) compute the mean and standard deviation, and compare the mean minus 5 standard deviations to the lower specification, and the mean plus 5 standard deviations to the upper specification. A capable process has the mean +/- 5 standard deviations at or within the specifications. In any case, the mean and standard deviations can be used for X-bar and R charts. (6) If the data *do* exhibit peculiar behavior, the sources of the behavior should be eliminated or reduced in their effect, unless it is inconsequential. X-bar and R-charts may not be appropriate for data dominated by peculiar behavior (the mean and/or standard deviation are not appropriate).

Regardless of the outcome with respect to control charts, even an informal process capability study can be tremendously useful in tracking down sources of variability. An industrial statistician can take the study one step further and perform variance components designed experiments—to quantify the amount of variability introduced by various sources.

5.5.3   Interpretation, and Resultant Action

X-bar Charts

X-bar values are averages, hence follow the behavior of the overall level of the underlying parent distribution. If the parent distribution shifts up or down in value, the X-bar values will do the same, and by the same amount. This type of shifting is precisely what we want to detect in many manufacturing situations; it is usually an indication of a process that is out of control. The process is being affected by a special cause of variation which is changing the level of the product characteristic being charted. Such special causes might be any of the following: changes in material, operator, inspector, set-up, SOPs, machine setting, supplier; calibration of a measurement system, wear of a tool, knob-twiddling, humidity, overheating/cooling, human error. Anyone with experience in manufacturing can easily add to this list.

Some shifts that are seen in X-bar charts are not due to the type of out-of-control behavior we are trying to identify and correct. They include: (1) outliers in the data—which are consequences of a machine or human transcription error, not an out-of-control process; (2) change in proportion of mixture distribution components—if the parent distribution has components at different levels, then a change in component proportions may bring about a change in average level. An example of the latter is varying proportions of alloys 1 and 2 in the mix of steel alloys used to manufacture ball bearings. Inventory or delivery schedules could result in a change in proportions from, say, 3:1 to 1:3, which could trigger an X-bar out-of-control signal if one of the alloys makes smaller (or larger) bearings than does the other.

Other shifts in level which may show up in an X-bar chart are due to behavior more appropriately tracked and understood through the use of the R-chart. For example, a change in variability is specifically what the R-chart is intended to identify, but an increase in variability can lead to very high and very low x values—hence, very high and very low X-bar values. A decrease in variability can lead to the center-hugging phenomenon which is sometimes observed in X-bar charts.

R-charts

R values are ranges, so a change (particularly an increase) in an R value which triggers an out-of-control is due to a change in the variability of the data, measured group by group in time order. Most increases in variability are due to the introduction of a new exacerbating source of variability which is inconsistent with similar parts of the process (such as a new fixture which is similar to, but not quite the same as other fixtures in use), or the failure of a controlling process which ensures consistency (such as a thermostat on soldering preheat). Other special causes of variability resulting in a change in range might include, amoung others: replacement of trained operators with inexperienced operators, non-uniform material, new equipment, and overadjustment of a process. Those causes listed as possible culprits in X-bar out-of-control situations can also apply.

The R-chart can, as well, be sensitive to the same circumstances as might inappropriately trigger an X-bar chart out-of-control: outliers or change in proportion of mixture distribution components. These causes are harder to identify.

P-charts

P values are proportion defective, so a shift upward from the in-control level of p values means nothing more than that there is a higher proportion of defectives. An out-of-control indication on a p-chart is a weak signal compared to an out of control on an X-bar or R-chart because only binary data are being used. It simply means that a failure mode has become active which causes such a high proportion of reject that it cannot be explained by the normal variation in the in-control distribution of p values. Special causes responsible for shifts upward in p values are essentially the same as those responsible for shifts in either direction on X-bar charts, plus those responsible for shifts upward on R-charts.

Though it is true of users of all control charts, it is especially incumbent upon the users of p charts to carefully define what is meant by defective, and ensure consistency of the application of the definition. It is also important to identify failure modes of the equipment, or person, or process which is being charted, along with the expected manifestations of the failure modes on the charts.

5.5.4 Management Commitment

Successful use of control charts in manufacturing depends upon more than just satisfying the technical assumptions, though this is important. Other conditions include a top-down commitment in the plant to their use, training in how to make them and interpret them, a change in SOPs and manager's expectations so that people at all levels realize the need to use them, and the willingness of management to visibly take action on the basis of control chart results—even if it means shutting down a line, shutting down a piece of equipment, rejecting a vendor, or decertifying an assembly line worker from performing a certain task.

Top-down commitment to anything new is a much bally-hooed and rarely obtained commodity. If top management were to commit to even a fraction of what they are asked to, they would, first, be overcommitting their personnel, and, second, be implementing programs that might well take opposing means to achieve the same ends. Getting top management to commit to control charting is not simply asking for commitment to the use of an isolated and helpful tool which will increase quality, increase yields, decrease costs, and so on. Rather, it involves a change in outlook toward manufacturing processes to include the statistical process engineer's perspective. For the statistical process engineer, control charting is one of several key tools that help him in his quest: to understand the variability of a process, to decompose the variability into its components, to narrow down and focus on reducing or eliminating those most important components, then to control and constantly improve the process. Until top management understands the nature of variability, it cannot move completely into the realm of objective decision-making.

Objectivity in making decisions is worth a lot to managers because it lets them take or reject risks knowingly. Without understanding variability, managers have to pay a costly insurance

premium; managers take smaller risks than desired, due to uncer-
tainty in the *true* level of risk. Such decisions, though perhaps not
biased one way or the other, will necessarily be sometimes capri-
cious. Managers will occasionally overreact to merely random fluctuations both in accounting data—such as yields, defect rates, number produced—and in product or process measurement data, such as
kiln temperature or tensile strength.

Thorough training in how to make and use control charts is
needed, for obvious reasons. It should be on at least three levels:
(1) conceptual, for top managment, (2) technical, for middle-
management technical people, and (3) cookbookish with a little
conceptual content, for operators. An external consultant, or at
least someone from another part of the company, is usually most
effective in initial training, because he can bring authority and will
seem to have no political axe to grind. He must be committed for
the long-term, however, to see the project through to implementation. Otherwise, control charting will likely die off as just another
management fac. The trainer should be strong in statistics. It is
wise to place the conceptual emphasis of the training on understanding variability more than understanding the concepts of different
control charts (x-bar, r, s, p, c, u, etc.). The cliche " actions speak
louder than words" seems to hold very strongly here. This author
has repeatedly received a cool reception to illustrative but fabri-
cated examples of control charts, but as soon as a control chart of
real and recent data is shown, the incentive to understand multi-
plies. Fruitful and lively discussions follows. Lastly, training
should be taken seriously, with high-quality educational materials.
The trainer should be held professionally responsible for good
education.

Changing standard operating procedures (SOPs) in the plant is
one of the hardest steps of statistical process control for several
reaons. People become set in their ways. Some people are afraid
that they might be replaced by a computer. Some harbor professional resentment. It is not effective to simply add on the responsibility of making control charts, interpreting them, and acting
on them to the other responsibilities of an operator or technical
person; something has to be given up. Otherwise, there will be
bad attitudes, a feeling of overcommitment, or shortchanging of

other tasks, as well as transitional mistakes. Plus, those who have not been trained, but who are being subjected to judgment by a chart they cannot read, may be demoralized or feel that they are losing turf.

Unfortunately, the risk of being replaced by a computer is real. Let us face the facts. Control charts can automate and improve efficiency of both parameter-sensitive processes (for example, preheat temperature in soldering, or mixture of material in a chemical process), and of equipment maintenance. Fewer manufacturing people are often needed, but they must be more technically sophisticated. This is the job of management, IR, and perhaps ultimately our nation's parents and schools. People who do not embrace statisticial process engineering techniques can be resentful, because these techniques implicitly label the old way of doing things *inadequate*, which it is. This attitude can be mixed with attitudes of stubbornness and paranoia.

5.5.5 Ultimate Objective: Reduce Product Variability

To close this section, a few philosophical words are given to remind the reader that the purpose of statistical process control, and using control charts within an institutionalized control loop, is to *reduce product variability, and meet specification targets.* This is one of the chief measures of quality, which leads to customer satisfaction: product consistency. It is a hallmark of many Japanese success stories. A FIS can help to bring it about by institutionalizing and at least partially automating control loops.

5.6 WHAT STATISTICAL CONTROL CHARTING WILL AND WILL NOT PROVIDE FOR YOU

Control charts *will provide* several valuable things:

1. Graphical presentation of data, with limits

2. A way to monitor and react to measured process or product values, and counts of defects or defectives

3. A theoretically sound and practical procedure which can be used with any desired level of confidence to declare out-of-control (this is traded off with confidence that a true out-of-control situation will be detected)

4. Almost any desired length of trend detection

5. Objective (*not* arbitrary) exception reporting

Control charts *will not provide:*

1. A quick fix

2. Infallible problem detection

3. Suggestions on how to solve problems

4. A universally or eternally applicable tool

5. A substitute for engineering judgment

5.7 PATTERN RECOGNITION AND INTERPRETATION OF X-BAR, R, AND P CHARTS

This section provides a compendium of sample charts which illustrate patterns frequently seen in X-bar, R, and P charts. The selection of patterns is intended to comprehensively cover singly occurring patterns, such as would result from a single mechanism causing a process to depart from a state of control. It does not include possible combinations of two or more patterns which might result from one or more underlying mechanisms causing a process to depart from a state of control. See the *Western Electric Handbook* for further discussion of some of these patterns.

5.7.1 Pattern Recognition and Interpretation of X-bar and R Charts

1a/b. Extreme and sudden shift in an X-bar / R Chart

2a/b. Slight and greatly-persisting shift in an X-bar / R Chart

3a/b. Moderate shift of moderate duration in an X-bar / R Chart

Sample charts begin on the following page.

1a. Extreme and sudden shift in an X-bar chart

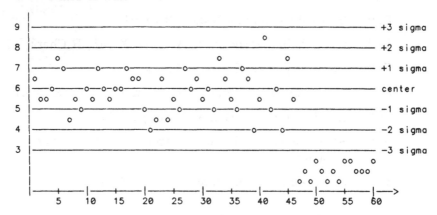

Sudden change in material, operator, lighting, inspector, set-up, SOPs, machine program/
    mode/setting, fixture, measurement system
Knob-twiddling
Human error
Electro-static discharge
Catastrophic failure of equipment, process, operator, etc.

1b. Extreme and sudden shift in an R-chart

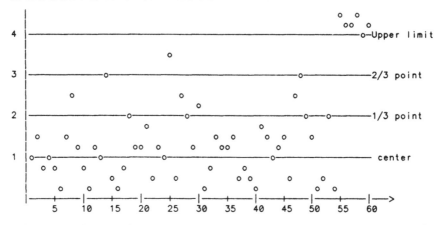

Sudden change in material, operator, lighting, inspector, set-up, SOPs, machine program/
    mode/setting, fixture, measurement system
Knob-twiddling
Accidental damage
Subtraction error
Catastrophic failure of equipment, process, operator, etc.

2a. Slight and greatly-persisting shift in an X-bar chart

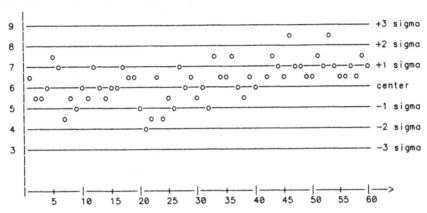

Gradual change in material, operator, lighting, inspector, set-up, SOPs, machine program/
   mode/setting, fixture, measurement system
Gradual wear of tools, threads, fixtures, gages, bearings, etc.
Gradual seasonal effects, such as fatigue, humidity, temperature, build-up of contamina-
   tion
Inadequate maintanance, bad housekeeping
Expansion/contraction

2b. Slight and greatly-persisting shift in an R-chart

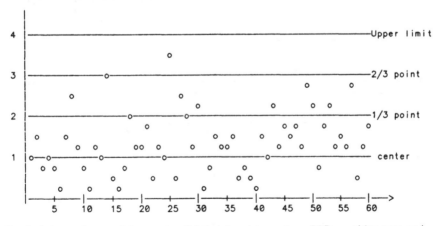

Gradual change in material, operator, lighting, inspector, set-up, SOPs, machine program/
   mode/setting, fixture, measurement system
Gradual wear of tools, threads, fixtures, gages, bearings, etc.
Inadequate maintanance, bad housekeeping

3a. Moderate shift of moderate duration in an X-bar chart

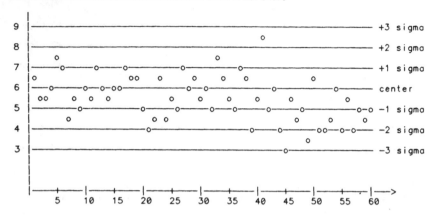

(See both 1a and 2a)

3b. Moderate shift of moderate duration in an R-chart

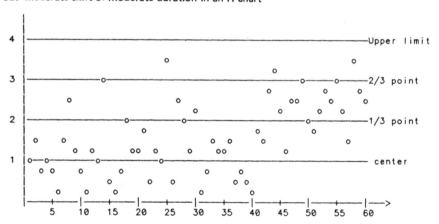

(See both 1b and 2b)

4a. Outliers in an X-bar chart

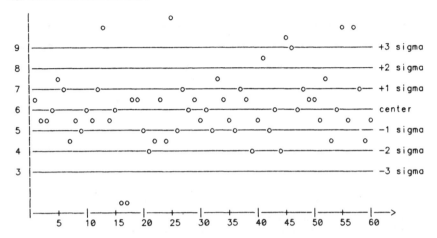

(Take note of R-chart, since an outlier usually causes an unusually large range value,
   otherwise, values may have moved in a group)
(See 1a)
Human error
Measurement system failure
Heterogeneous sample population (comparing a few apples with a lot of oranges)
Machine or operator or process is catastrophe-prone or has transient failure-modes

4b. Outliers in an R-chart

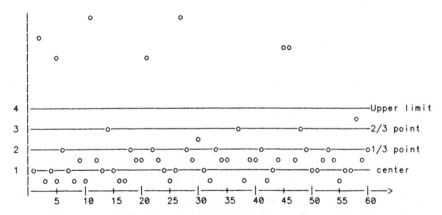

Accidental damage
Subtraction error
Occasionally heterogeneous sample population (comparing a few apples with a lot of
   oranges)
Machine or operator or process is catastrophe-prone or has transient failure-modes

5a. Gradual trend in an X-bar chart

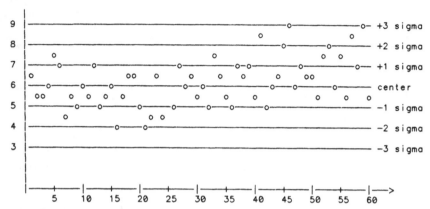

Wear of tools, threads, fixtures, gages, bearings, etc.
Gradual seasonal effects, such as fatigue, humidity, temperature, build-up of contami-
    nation.
Inadequate maintenance, bad housekeeping
New imbalance of schedules or rates of operation
Gradual change in material, operator, lighting, inspector, set-up, SOPs, machine program/
    mode/setting, fixture, measurement system
Expansion/contraction

5b. Gradual trend in an R-chart

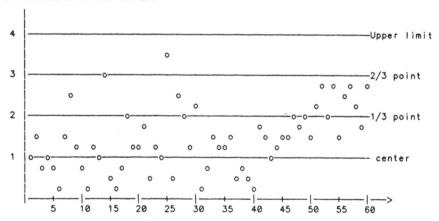

Wear of tools, threads, fixtures, gages, bearings, etc.
Gradual seasonal effects, such as fatigue, humidity, temperature, build-up of contami-
    nation
Inadequate maintenance, bad housekeeping
New imbalance of schedules or rates of operation
Gradual change in material, operator, lighting, inspector, set-up, SOPs, machine program/
    mode/setting, fixture, measurement system
Increasing problem of over-correction

6a. Cyclic behavior (periodicity) in an X-bar chart

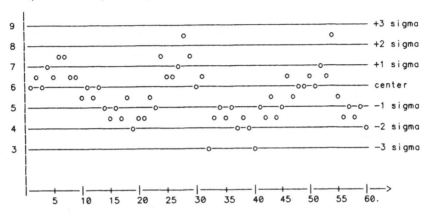

Seasonal effects, e.g., fatigue, humidity, temperature, contamination
Expansion/contraction
Shift change
Reflection of maintenance schedule (usually inadequate maintenance)
Voltage fluctuation
Worn threads or eccentricity on round or cylindrical parts
Under-correction or over-correction

6b. Cyclic behavior (periodicity) in an R-chart

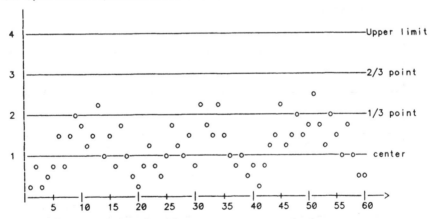

Seasonal effects, e.g., fatigue, humidity, temperature, contamination
Expansion/contraction
Rotation of fixtures, testers, gages
Shift change
Reflection of maintenance schedule (usually inadequate maintenance)
Voltage fluctuation
Too much play in fixture
Tool needs sharpening

7a. Instability in an X-bar chart

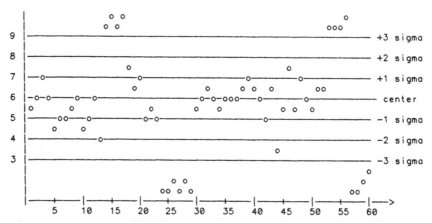

Over-control of a machine/process
Loose fixtures
Inappropriate substitution of material, operator, inspector, process, measurement system
Mixture of materials/operators/inspectors/processes/measurement, most of which are
   alike
Careless operator
Deliberate "running on the low/high side" of specifications

7b. Instability in an R-chart

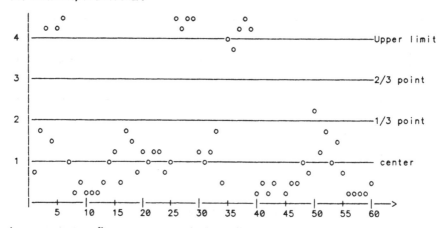

Loose contacts or fixtures, or too much play in fixtures
Inappropriate substitution of material, operator, inspector, process, measurement system
Mixture of materials/operators/inspectors/processes/measurement, most of which are
   alike
Careless operator

8a. Local bunching or random-walk in an X-bar chart

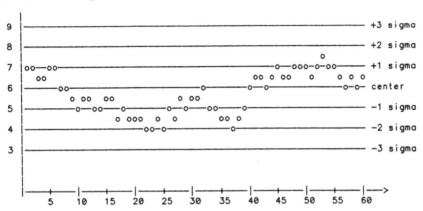

Human memory influencing measurement error
Gradual change in material, operator, lighting, inspector, set-up, SOPs, machine program/
    mode/setting, fixture, measurement system
Over-control
Knob-twiddling
Seasonal effects, e.g., fatigue, humidity, temperature, contamination

8b. Local bunching or random-walk in an R-chart

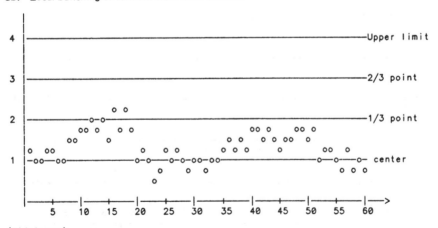

(This is rare)
Changes in the mixture of materials
Seasonal effects, e.g., fatigue, humidity, temperature, contamination

9a. Binning (long-term) in an X-bar chart

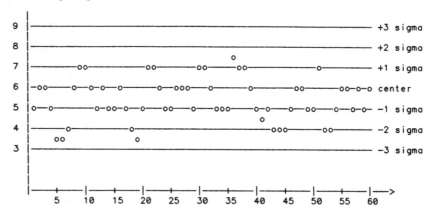

Measurement system lack of precision
Computer round-off or intentional grouping
Human preference for certain values, such as 2.5 over 2.3, 3.80 over 3.79, and for 0 over
    almost anything
Graph paper or plotting resolution is too coarse
Inadequate plotting capability or skill
Discrete settings on equipment, fixture
Mixture of materials/operators/inspectors/processes/measurement systems which are at
    different levels

9b. Binning (long-term) in an R-chart

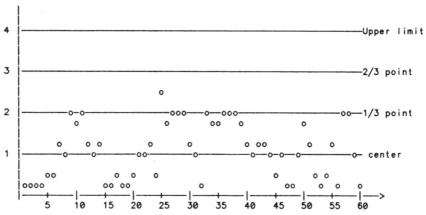

(See 9a)
Rounding error during subtraction to get range

10a. See-sawing (negative auto-correlation) in an X-bar chart

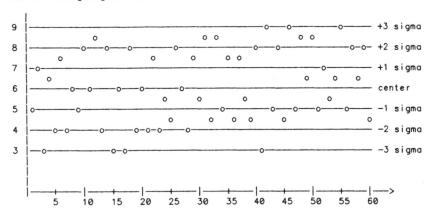

Over-control of a machine/process
Loose fixtures
Fast (relative to sampling frequency) seasonal effects, such as fatigue, humidity, temper-
    ature, contamination

10b. See-sawing (negative auto-correlation) in an R-chart

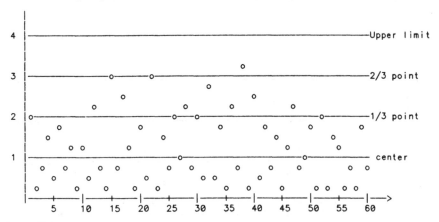

Loose fixtures
Fast (relative to sampling frequency) seasonal effects, such as fatigue, humidity, temper-
    ature, contamination

11a. Center-line-hugging in an X-bar chart

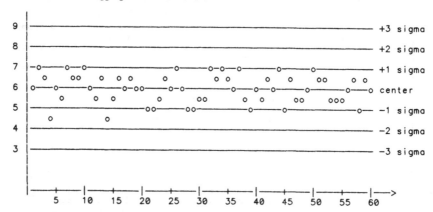

Control limits are obsolete (process has been brought under tighter control)
Control limits are incorrectly plotted
A former mixture parent distribution has become pure
Data are being prescreened for values moderately deviant from the average
The product is being prescreened before these measurements are taken

11b. Center-line-hugging in an R-chart

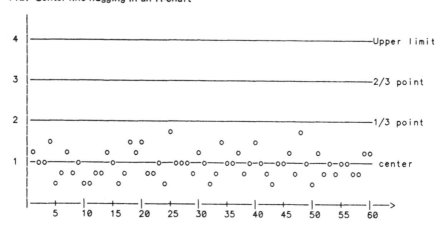

Control limits are obsolete (process has been brought under tighter control)
Control limits are incorrectly plotted
A former mixture parent distribution has become pure
Data are being prescreened for values moderately deviant from the average
The produce is being prescreened before these measurements are taken

5.7.2   Pattern Recognition and Interpretation of P Charts

      1c. Extreme and sudden shift in a P-chart

      2c. Slight and greatly-persisting shift in a P-chart

      3c. Moderate shift of moderate duration in a P-chart

      4c. Outliers in a P-chart

      5c. Gradual trend in a P-chart

      6c. Cyclic behavior (periodicity) in a P-chart

      7c. Instability in a P-chart

      8c. Local bunching or random-walk in a P-chart

      9c. Binning (long-term) in a P-chart

     10c. See-sawing (negative auto-correlation) in a P-chart

     11c. Center-line-hugging in a P-chart

Sample charts begin on the following page.

1c. Extreme and sudden shift in a P-chart

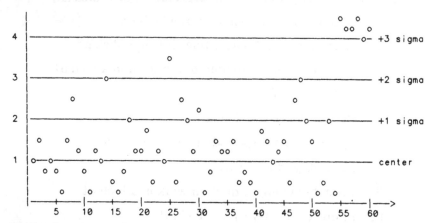

Clogging, jamming, part falling off, misalignment of fixture, bad contact, wrong part
Change in material, operator, lighting, inspector, set-up, SOPs, machine program/mode/
    setting, fixture, measurement system
Knob-twiddling
Accidental damage
Catastrophic failure of equipment, process, operator, etc.

2c. Slight and greatly-persisting shift in a P-chart

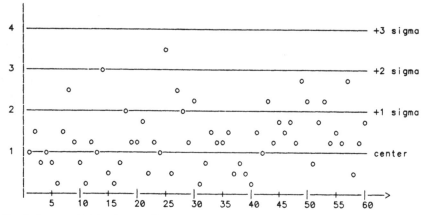

Gradual wear of tools, threads, fixtures, gages, bearings, etc.
Gradual seasonal effects, such as fatigue, humidity, temperature, build-up of contamina-
    tion
Inadequate maintenance, bad housekeeping
Expansion/contraction

3c. Moderate shift of moderate duration in a P-chart

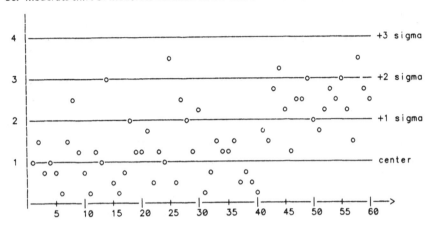

(See both 1c and 2c)

4c. Outliers in a P-chart

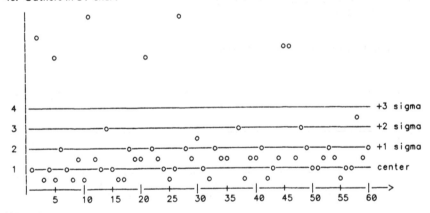

Human error
Measurement system failure
Heterogeneous sample population (comparing a few apples with a lot of oranges)
Machine/operator/process is catastrophe-prone
(see 1c)

5c. Gradual trend in a P-chart

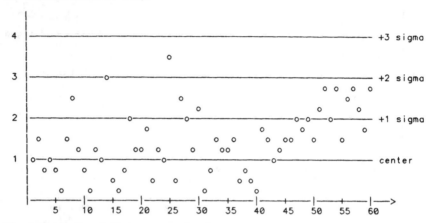

Wear of tools, threads, fixtures, gages, bearings, etc.
Gradual change in material, operator, lighting, inspector, set-up, SOPs, machine program/
    mode/setting, fixture, measurement system
Inadequate maintenance, bad housekeeping
Gradual change in mix of material

6c. Cyclic behavior (periodicity) in a P-chart

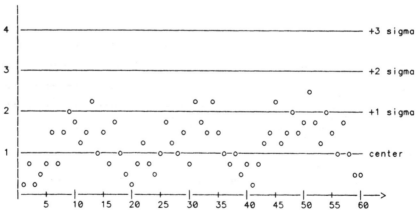

Fast seasonal effects, e.g., fatigue, humidity, temperature, contamination
Expansion/contraction
Shift change
Reflection of maintenance schedule (usually inadequate maintenance)
Worn threads or eccentricity on round or cylindrical parts
Under-correction or over-correction

7c. Instability in a P-chart

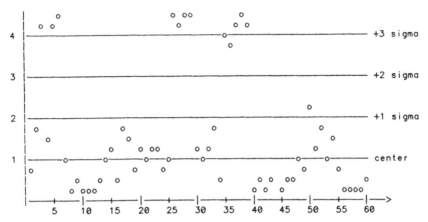

Over-control of a machine/process
Loose fixtures
Inappropriate substitution of material, operator, inspector, process, measurement system
Mixture of material/operators/inspectors/processes/measurement, most of which are alike
Careless operator
Deliberate "running on the low/high side" of specifications

8c. Local bunching or random-walk in a P-chart

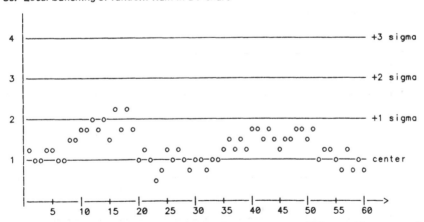

Gradual change in material, operator, lighting, inspector, set-up, SOPs, machine program/mode/setting, fixture, measurement system
Seasonal effects, e.g., fatigue, humidity, temperature, contamination

9c. Binning (long-term) in a P-chart

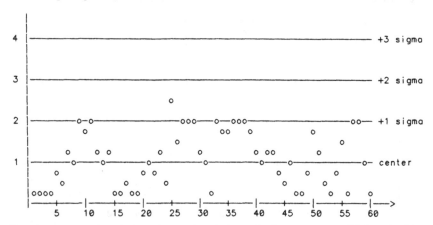

Pns, the product of typical p values and typical sample sizes are too small (they should
 be at least 5)
Measurement system lack of precision
Computer round-off or intentional grouping
Graph paper or plotting resolution is too coarse
Inadequate plotting capability or skill
Discrete settings on equipment, fixture
Mixture of materials/operators/inspectors/processes/measurement which are at different
 levels

10c. See-sawing (negative auto-correlation) in a P-chart

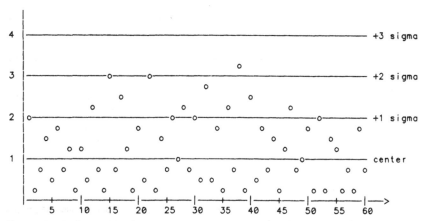

Over-control of a machine/process
Loose fixtures
Fast (relative to sampling frequency) seasonal effects, such as fatigue, humidity, tempera-
 ture, contamination

11c. Center-line-hugging in a P-chart

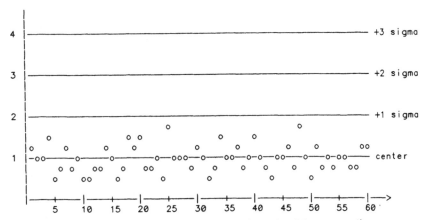

Control limits are obsolete (process has been brought under tighter control)
Control limits are incorrectly plotted
A former mixture parent distribution has become pure
Data are being prescreened for values moderately deviant from the average
The product is being prescreened before these measurements are taken

6

Decision Support and Automation Using Artificial Intelligence

6.1 INTRODUCTION

The previous Chapter 5 reviewed how data continuously collected from a manufacturing operation can be statistically analyzed to determine if a significant deviation from normal fluctuation occurs. If significant deviation is detected a warning can be automatically issued which initiates some action toward the cognizance, diagnosis, and correction of the cause of deviation. Statistical analysis thus provides a means of detecting out-of-control situations but cannot, of itself, provide any diagnosis leading to corrective action. Statistical analysis is also unable to deal with imprecise situations caused by ill-defined concepts and incomplete data. This must be done by a more intelligent agent—a person or possibly a form of computer programming that can do simple reasoning.

Today a very tenuous and as yet undeveloped way of dealing
with these imprecise situations is provided by the attributes of the
various Artificial Intelligence (AI) programming languages and
techniques. AI programs can potentially diagnose a situation in
which some data is missing and propose or directly control the actions required for correction. The AI programs are composed of
facts (assertions) and statements which relate the facts to each
other. Unlike conventional structured programming languages
that follow specifically predefined paths to foregone conclusions,
these AI programs can execute by matching the assertions to
queries and search to find answers in much the same way humans
think.

A common attribute of most AI programs is that they are
symbolic rather than numeric. This results in a higher degree of
abstractness, making the technique more difficult to understand
but potentially very powerful. This potential was tested a few
years ago when AI was directed to such problems as the diagnosis
of human diseases. Because of the extreme complexity of these
problems and issues related to the social acceptance of such a personal and emotional application, success has been marginal.

Today, successfully applied AI programs are written by
those who fully understand a well-defined application domain
such as manufacturing. To acquire this understanding the professional AI computer scientist who creates useful programs for
decision support and automation needs to be tightly coupled to
the manufacturing application expert.

It is not necessary to use AI languages to write AI programs.
Any software programming language that can do conditional
branching as illustrated in Figure 3.12 of Chapter 3 can be used
to write rule based programs that progress through a decision tree
to arrive at a solution. This is illustrated in Figure 6.1. The solution will be as accurate and precise as is the applicability of the
decision conditions programmed for each node or branch point
of the tree.

What AI languages and programming tools do is provide a
much easier environment in which to write this type of program.
This means that prototype diagnostic and decision support systems can be developed very rapidly and therefore at considerable
savings. The newer versions of AI languages produce programs
that execute fast enough, when properly structured, for factory

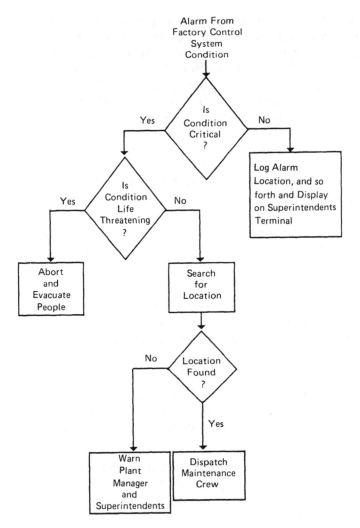

Figure 6.1 Typical decision tree.

trials to confirm the functionality of a new system. They can often be used directly for final system implementation or portions can be rewritten in a more basic structured language like C or PASCAL to further reduce the execution time. AI languages, therefore, provide a new tool for applying computers to industry. They are potentially the next generation of computer languages to follow the structured languages described in Chapter 3.

This chapter endeavors to describe AI from an applied viewpoint. A glossary is included at the end of the chapter to define some of the copious AI jargon inherent with the field.

There are many ways to define AI. Perhaps the best way is to bracket these many definitions by selecting a few examples.

1. AI is a branch of computer science concerned with the study of knowledge representation, search, and reasoning in problem solving.

2. AI is the act of applying modern computation to a situation that is sensed through some means of perception or detection such that the results of the computation produces some controlled action or movement.

3. AI is the application of computers to sense our environment and perform tasks done by humans. These include diagnostic analysis, design support, decision support, vision, tactile sensing, motion control, computing, programming, learning, speech recognition, and language translation.

4. AI is a software and hardware tool, resulting from the natural evolution of computer science, that can be applied to many activities including manufacturing control problems not easily defined in specific quantifiable terms. The software thus allows non-determinant situations and approximations to be programmed and controlled. (The author happens to like this definition because he believes it is realistically achievable!)

The decisions provided by AI programs are not absolutely reliable or accurate (as is also the case with human experts). This is because AI can deal with problems where some of the know-

ledge needed to make a completely accurate and reliable decision
may be missing. In this context, AI is dealing with real situations
where intuitive judgment, inherent with the combination of facts
in the knowledge base, can be as powerful a factor as concise data.
Real world problems are boundless when the extremes of solution
reliability and accuracy are considered. AI deals with this bound-
lessness by possessing an attribute that as more knowledge is programmed into a system the more reliable and accurate are the solutions suggested by such a system.

6.1.1 History of AI

The grandfather of AI was the English philosopher Thomas Hob-
bles whose life and ideas are discussed by Haugeland (1985). In
the 1650s his concept of thinking in terms of symbolic representation formed the basis of todays' AI technology. The use of
symbolic representation and methodicial rules made the thinking
process clear and rational. The first references to AI appear around
1842 by those who wrote about Charles Babbage, England's
eccentric and first computer scientist. He contributed the concepts
of programmable operations and conditional program branches. It
then took a hundred years for the vacuum tube, transistor and
integrated circuit to be invented and developed as building blocks
for electronic computers. The name, Artificial Intelligence, coined
by John McCarthy did not appear until 1956. In 1958 he and
Marvin Minsky formed the AI Laboratory at the Massachusetts
Insitute of Technology. The next year he invented a new kind of
computer language for processing lists named LISP. Common LISP
has become the current language of choice for general AI applications. It is interesting to note that FORTRAN was developed at
about the same time!

The late 1960s was the dark age for AI with very little measurable results except for the proposal of the "resolution" inference rule. In 1973 a renaissance began with the writing of the
first modern AI programs. Another burst of creativity occurred
beginning in the late 1970s. At this time the AI language PRO-
LOG migrated to the United States from France and Scotland.
The Japanese adopted PROLOG as the language of their fifth generation computer project in 1980. By this time computing machines designed to support AI programming were available at a cost

of about one million dollars! In 1985 the price of these machines had dropped to the range of 5 to 100 thousand dollars, multiple programming shells were available on each machine and many people had been trained to apply AI. Progress from 1982 to 1987 has come more from advances in integrated circuit and computer systems architecture than from fundamental advances in AI. The future will probably be strongly shaped by what is learned from AI's growing application to domains such as manufacturing.

6.1.2 AI Application Areas

There are six general areas where AI is applied. These areas are:

1. Diagnositcs

2. Robotics

3. Perception

4. Computer Science

5. Learning

6. Communications

Of greatest interest to manufacturing people are diagnostic and robotic applications.
     One of the most popular methods of applying AI to diagnostic applications is in the form of an Expert System because it is well-suited to decision making. Expert systems are used to assist in the diagnosis of situations that are either very complex, require very rapid solution or a combination of both. These systems address the issues of repair, maintenance, configuration, control, and design. A description of the Expert System follows later in this chapter.
     Robot systems use both conventional and AI programs to manage the sensing of the robots' environment, to control the motion of the robot, and to instruct the robot for new tasks. Expert systems, running in real time, are used for robotic cell diagnostics.
     Computer science is being served by its AI prodigy through the development of support systems for programming and debugging. AI is also creating incentives for new types of computing

architecture and hardware that better support AI applications. Prominent examples are the use of parallel architectures for vision system image processing, computing machines designed to support a specific AI language, and the fifth generation computer systems currently being developed by the Japanese.

The basic art of learning is part of the AI repertoire. Extensive research is being conducted to discover ways computers can monitor and learn from situations. As results accrue from this work, manufacturing will be one of the first activities to benefit. Computerized manufacturing control systems need to be maintained in a current state of knowledge about the parameters that define semi-static conditions on the shop floor. As these conditions change, people have manually changed existing computer files and added new files. The development of AI-based learning systems, used to automatically up-date these parameter files, will save considerable labor and maintain timely control algorithms.

There is considerable AI effort expended on natural language translation, generation, and recognition. Voice input systems have been used by industry since the early 1970s for inspection, monitoring, and control applications where human eyes, arms, and feet are otherwise fully occupied. Example applications are the visual inspection of integrated and hybrid circuits, the recording of repair diagnostic information, and package routing at truck transport exchanges. As integrated circuits become available to support these voice recognition systems, cost will further decrease permitting their widespread use for efficient data input by factory personnel.

6.2. KNOWLEDGE BASE (OR EXPERT) SYSTEMS

Decision support systems represent one of the greatest practical applications of AI. As such they are of interest to manufacturing personnel. These are called Knowledge Based or Expert systems. In this section a type of Knowledge Based system will be described and some concepts of AI illustrated by a simple PRO-LOG example.

The Knowledge Based system (KBS) provides a way to combine the knowledge of several human experts into a single tool that can be used by less experienced persons to assist in the solution of problems and the control of situations. The advantages are obvious. A disciplined method is provided for problem analysis with the potential of growing to become more powerful than any single expert. The system reduces the often drastic loss of expertise when an expert leaves or must be reassigned. There are problems, however, which are still being solved. One is the inconsistency in knowledge that is acquired from different sources. Another is the slow and difficult acquisition by the KBS of this knowledge. A third limitation is the lack of generality suffered by these system which make them very domain dependent.

The KBS is composed of the basic parts shown in Figure 6.2. The knowledge base contains information related to the specific domain or problem to which the expert system is applied. The knowledge base serves the same function as the data base used in conventional computer systems but in addition can include information about logical relationships and possibly their reasons for existing.

The inference engine is the mechanism used to control the processing of the knowledge to arrive at the desired result. It is an internal and intrinsic part of the AI environment and is composed of a set of high-level programs that define the method used for problem solving.

There are two basic methods used by these systems to solve problems. The methods are called goal driven or backward reasoning and data driven or forward reasoning. In goal driven systems the search begins with the goal to be proven and attempts to satisfy all the conditions that would make the goal true. In the data driven system the search begins with everything known about a problem and attempts to draw conclusions based on the data. The example which follows is a goal driven form of KBS.

This KBS works by searching the knowledge base, under the control of the inference engine, for inferences (logical conclusions) that apply to the problem posed by the user. The logic and facts used to make these conclusions are contained in the knowledge base. The methods for searching the knowledge base are contained in the inference engine and in the rules of the knowledge base which are programmer dependent. The techniques used to match the posed problems to the contents of the knowledge base are contained in the inference engine. To infer is to derive a conclusion from facts and premises. This type of KBS tests the facts in its knowledge base against a premise or statement of fact that is

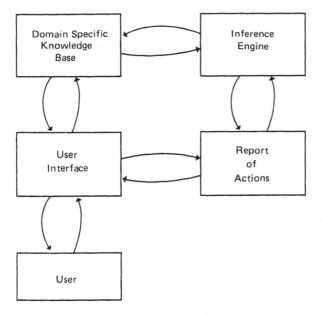

Figure 6.2 The basic expert system.

introduced as a possible solution to the problem under consideration. If the premise proves to be true, it becomes a solution to the
problem.

    The testing of the theorem is accomplished by a search tech-
nique called backward reasoning or backward chaining. Often the
searches are limited by the person who writes the program because
of heuristic (experience) knowledge. This can greatly reduce execution time and takes advantage of both the intuitive knowledge
and specific knowledge humans have about the problem.

6.2.1 A PROLOG Example

This example is chosen because PROLOG is an excellent backward
reasoning theorem proving language for writing expert systems and
because the concept of AI is easier to demonstrate using PROLOG
than with other AI languages. The objective is to give the reader
a sense of how the programs work and how facts and relationships
can be added to an existing AI system to make it perform new

tasks and execute old tasks with greater accuracy and reliability.

PROLOG is an AI programming language that simplifies informing a computer about obvious and not so obvious facts. The name is a contraction of "PROgramming in LOGic." The approach is to describe the facts and rules about a situation with a series of assertions or statements. These assertions constitute the knowledge base. Program execution is controlled by user defined rules in the assertions and fixed rules stored in the inference engine. The program tries in every way contained in the rules and inference engine to prove each proposition posed to the program to be true or false. Thus the PROLOG program is a knowledge base of logical assertions regarding facts and relational rules describing a situation.

Queries about a specific proposition associated with the situation initiates a systematic search of this knowledge base for a statement that justifies the proposition. The search is "depth first." This means the search proceeds down a tree-like structure of knowledge organization to the bottom of the first branch and then returns to the next branch to repeat the same process until a matching solution is found or all the branches are traversed.

If a statement is found that justifies the proposition, it forms the solution to the query. If none is found it is assumed that the answer cannot be derived from the available knowledge. From the above explanation it is apparent that as additional knowledge enters the system, problems that previously could not be solved become tractable provided the knowledge is consistent.

There are many versions of the PROLOG language, each using slightly different notation. For this reason the reader should anticipate having to adapt to different language symbols. Fortunately the basic components of the language remain consistent as defined below:

Terms: Usually lower case characters or words that have fixed values such as "a", "b", "table", and "john".

Predicates: Describe relationships such as "on", "mother", and "friend_of".

Variables: Usually upper case letters or words preceded by a symbol such as "%who" or "%X". They represent a quantity or thing that has not been fixed or determined. They are used in the same way variables such as "x" and "y" are used in algebra.

Propositions: These are assertions or statements of fact unless followed by a symbol such as "?". These symbols are language dependent. When used with the symbol they are queries or theorems to be proved correct. An example is "on (p, table)", which means "p" is "on" the "table".
An example of a query is, "father (%who, bill)?". This means "who" is the father of "bill"?

The propositions and predicates use connectives to combine statements and/or relationships. A set of PROLOG symbols for these connectives are shown in Table 6.1.

With the above introduction we can move to a simple example. It is especially helpful because of the abstract nature of AI to follow an example as an aid to understanding some of the concepts. Without an example the generalities become verbose. This simple case will deal with assemblies and subassemblies of a product, such as a television set, PBX telephone switch, or automobile. The example is very simple, containing the knowledge that the product is made of assemblies and the assemblies are made of subassemblies. With a short three line PROLOG program we will see how, in contrast to other programming schemes, many facets of the composition of the product can be determined.

Table 6.1 PROLOG Connectives

English	PROLOG Symbol
And	,
Or	;
Implies	<--
Not	not

The relationship between an assembly and a subassembly is expressed in PROLOG by the program assertion:

(1) compose(assembly, subassembly).

The subject, "assembly" of the relationship "compose" precedes the object of the relationship, "subassembly". The assertion states that assemblies are composed of subassemblies.

One can now ask the above single line PROLOG program whether assemblies are composed of subassemblies. The proposition for this query takes the form:

(2) compose(assembly, subassembly)?

The response generated by PROLOG's inference engine is:

(3) yes

because an exact match was found between the query (2) and the program assertion (1).

If the proposition had been:

(4) compose(assembly, doughnuts)?

The answer would be no because the program does not have any information about doughnuts. Thus we see that the PROLOG language itself supplies analysis techniques and responses that must be specifically written if conventional structured languages like PASCAL and C are used.

More complicated queries can be asked by using variables. For example, "What are assemblies composed of"?

(5) compose (assembly, %Z)?

The term assembly is the subject of compose. The unknown variable %Z is what the assembly is made of.

The program will search its knowledge base (PROLOG program statements) to find any assertions that match the above query. If it finds a match, the inference engine will assign values to the variables that are consistent with the match and present them as a solution to the query. In this case the result will appear as:

(6) %Z = subassembly

because this matches the assertion in the knowledge base:

(1) compose (assembly, subassembly).

If no match were found the result would be a no. If additional statements about the composition of assemblies had been in the knowledge base, the entire list of compositions would be produced.

A second assertion related to assembly composition can be added to the beginning of the knowledge base. The PROLOG program would then appear as:

KNOWLEDGE BASE

(7) compose (product, assembly).

(1) compose (assembly, subassembly).

People can easily deduce from the above two assertions that products, made from assemblies, are therefore made from subassemblies even though this is not specifically defined by the assertions. PROLOG can also make this deduction if the knowledge base has information that supports this relationship. An AI professional will write relational information in a very general form without uniquely identifying the assemblies. In this way the general information assertion can be used for any kind of assembly-subassembly composition. The assertion simply states that entities are composed of many intermediate levels of other entities. To add this information to the knowledge base, so that three levels of assemblies can be recognized, a third assertion is appended:

KNOWLEDGE BASE

(7) compose (product, assembly).

(1) compose (assembly, subassembly).

(8) made_from (%X, %Z) $<--$ compose (%X, %Y),
     compose (%Y, %Z).

This last assertion says that if X is made from Z then it is implied or true that X is composed of Y and Y is composed of Z. In this assertion X, Y, and Z are variables so this is a general case applicable to any situation where three levels of composition exist. The expression "made_from" is connected with a "_" because most computers only recognize a single group of connected characters as a definition or command rather than separated groups.

To make the relationship meaningful to English speaking people the single connected word "made_from" is used. The symbol "<— —" means "is implied by" or "is true if" and the symbol "," separating the two assertions to the right of "<— —" means "and". This logical "and" says that both of the compose assertions have to be true at the same time for the left hand "made_ from" proposition to be correct.

With the above knowledge base other questions that have not been previously anticipated and therefore specifically answered can be queried. For example, "Are products made from subassemblies"? The form of the query is:

(9) made_from(product, subassembly)?

The PROLOG program will execute by searching for each "made_from" assertion. When the left side of line 8 of the knowledge base is found the inference engine sets:

%X = product

%Z = subassembly

by matching %X and %Y to the above query. The definition then becomes for this specific instance:

(10) made_from(product, subassembly) <— —

compose(product, %Y), compose (%Y, subassembly).

Line 10 says that products are made from subassemblies if products are composed of a something, %Y that in turn is composed of subassemblies. The knowledge base will be searched to determine if this something represented by the variable %Y exists. The first two assertions (7) and (1) of the knowledge base will lead to the conclusion that "%Y = assembly" and the answer will be "yes" to the question, "Are products made from subassemblies"? Note how different this is from conventional forward search of structured programs where the query would have had to be anticipated in advance, a very detailed analysis program developed to provide the answer, and the answer itself specifically programmed to be displayed if the analysis supported the conclusion.

One final illustration of the level of abstraction and therefore the generic nature of PROLOG is appropriate. The above knowledge base contains enough information to answer the general query, "Find the levels of components that are used to build products". The query is very abstract and takes the form, "What is it, (%A) that is made from something, (%B)"? The "it" is the variable %A and the something is the variable %B. The PROLOG expression is simply:

(11) made_from(%A, %B)?

The answer produced without any additional programming effort will be that products are made from subassemblies. The form of the answer will be:

%A = product

%B = subassembly

Before leaving this example it is noteworthy that in PROLOG there is no explicit distinction between input and output arguments.

6.2.2 The Knowledge Base

Hopefully the PROLOG example has given some insight into how at least one type of AI language works. This understanding can be enhanced by mentally dividing the expert system into two parts, the knowledge base and the inference engine. The knowledge base is part of the code of a PROLOG program. The inference engine is inherent with the PROLOG language and comes along with its purchase without the user having to write any functional code. In this section the knowledge base is described in more detail. The next section will comment on the inference engine.

The knowledge base can be organized in various ways to facilitate the representation, search, location, acquisition and processing of information. How the knowledge is represented is very important. Every representation technique emphasizes certain information about a concept and ignores other information. A good representation makes the right information available for the problem's solution. The most often used forms of representation are terms, rules, frames and semantic networks.

Terms

Terms are simple statements of fact in the appropriate form for the programming language used. A term is illustrated by statement 7 of the preceding PROLOG example.

Rules

Rules describe relationships. They are based on heuristic knowledge and usually organized in a hierarchy. Heuristic knowledge is
learned or discovered through experience by the use of some
method or device. It comes from the Greek word heuriskein,
meaning to discover, find, or learn. Most plant managers, as an
example, know how to effectively direct a manufacturing operation because they have learned through experience the best actions
to take under certain circumstances. This knowledge gained
through the method of experience is heuristic knowledge. Rules
based on the laws of physics can also be used by the same AI program. Systems using these fundamental physical and logical rules
"reason from first principles."

     Rule oriented knowledge bases usually contain from a few
hundred to many thousands of rules. The major difficulty with
these rule-based systems is the excessive number of rules that must
be defined for the system to work well in real situations. It is
also difficult to establish concise definitions for rules.

     A popular representation for rules is derived from a method-
ology developed by mathematicians for dealing with logic. It is
called predictate calculus. It can be used to represent facts and to
define qualitative relationships. The logic symbols used in this
representation are processed by the computer in much the same
way mathematical symbols for addition and multiplication are
processed by a calculator. Examples of symbolic logic are given
by Gustaso et al. (1973). In logic, predicate means to make an
expression of a proposition or assertion about a subject. It may be
helpful to think of the predicate of a sentence which expresses
what is said about the subject of the sentence. Predicate calculus
is thus a formal language used to represent assertions or rules with
a computer. One of the commercially available programming
languages based on predicate calculus is PROLOG. A typical rule is
statement 8 of the preceding PROLOG example.

     Rules can also be represented in less general form by conven-
tional branching statements. Statement 8 of the previous PRO-
LOG example can be written in structured English as:

If

        Products are made from assemblies

        And assemblies are made from subassemblies

    Then

        Products are made from subassemblies

    Else

        Products are not made from subassemblies

    Continue

The problem with this conventionally structured language is it specifically applies to only the exact words used and many additional assignments have to be made to produce a working computer program.

Frames

Frames are a way to represent knowledge about a temporary situation. They are often used to describe the state of a physical environment where things are moving or changing. Typical manufacturing examples are the status and location of parts, fixtures, and so on in a work cell or the location of product as it flows through the manufacturing processes or the status of a machines capacity, tools, availability and so forth.

Frames are used to address a very difficult knowledge access problem excellently described by Haugeland, p. 203, (1985). The problem is how to know the salient side effects about a situation without having to investigate and eliminate all other possibilities that might have occurred. The side effects for the most part will not be pertinent but can, based on what is currently happening, become very much a part of determining what to do next. The difficulty is in knowing how to ignore selectively almost everything an AI system knows in order to hone in on relevant facts without the enormous effort required to rule out alternatives.

An example is the occasional alarm triggered by a process out-of-control condition. What action to take when the alarm occurs depends not only on the nature of the alarm, which is easy enough to detect and handle, but on the current state of the process. The current state is much more difficult to ascertain. This is because processes are constantly changing and involve thousands of descriptive details. Take a specific manufacturing example of a flexible manufacturing machining process. During this process the predominant time is spent cutting metal. A small part of the cycle is devoted to a transfer mode where, in this example, vacuum is used by a robot to hold product while it is being transferred to another location in the work cell. If the power goes off, the robot can only hold the product for a few seconds before it will be dropped as the vacuum is lost. An abort action could be to open a valve to an auxiliary vacuum system but only if the process cycle was in the product transfer mode. Frames are used to keep track of the state of a process so that actions, such as the one cited, which depend on that state can be determined. Frames and other files are often up-dated as events occur with the use of small event actuated programs called daemons.

Frames are thus sets of entities which describe a situation. They can be sets of hypotheses which are single or in composite form. An example frame is:

Product
    made_from
        product:assembly
        assembly:subassembly
    made_with
        cell:A
        tool:100

The application of frames requires that the knowledge base contain a set of prototypical situations described by frames. These represent the many expected states of a situation. The user describes a particular situation. The AI program then tries to match this particular situation by inference to one or more of the set of prototypical situations in the knowledge base. If a match is found it does not have to be exact. The relevant prototype is identified and the portions of the frame that do not match are used as suggestions for further action by the user.

Semantic Networks

Semantics is the art of interpreting the meaning of symbols including symbols that reflect the relationship between other symbols.

Semantic networks are a way of representing knowledge, without any reference to how it is physically organized, in which there are owner and member relationships. In these networks members are subordinate to owners and inherit characteristics from them. An example of such a network is shown in Figure 6.3. The network is defined by using terms (symbols) which are related to each other. When illustrated the terms appear as descriptive words and the relationships appear as arrows.

Search Techniques

Searching the knowledge base is done by moving forward through the series of nodes in a decision tree or assertions toward a conclusion or by moving backward from the proposition to be proven in an attempt to find all the necessary conditions to prove the proposition. Forward searching or chaining is typical of rule based systems where the observable conditions are stated first. Each if-then-else or similar branching condition is implied by the structure

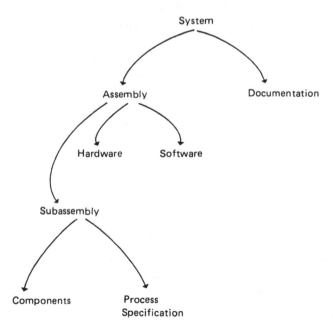

Figure 6.3 Illustration of a semantic network.

of the program. Execution is forwarded through these branching nodes until a conclusion is reached. This type of AI program can be written in any language that supports branching.

Backward searching or chaining is exemplified by the PRO-LOG approach illustrated earlier. Here the conclusions to be proved such as, made_from(product, subassembly)?, is stated first. The search is then conducted backward through the assertions until those necessary to prove the conclusion are found. One can appreciate that this type of search might involve a very large knowledge space and therefore consume vast computing resources and time before a solution or a no is returned to the user. To avoid this problem the search space is often limited by heuristics as described by Hadavi (1985). This technique provides practical solutions to otherwise intractable problems and yet maintains the inherent flexibility and knowledge growth capability of backward chaining systems.

The direction and type of search is controlled by the structure of the program and strategy being used. A breadth-first search progresses through each of the top levels of a hierarchy before dropping to the next set of levels. The strategy is thus similar to scanning a typical newspaper where the highlights are written first followed by greater and greater detail as the reader progresses through an article. A depth first search works its way to the bottom of the first branch of a search tree before returning to the top and progressively moving to the bottom of each succeeding branch. Often both forward and backward chaining and breadth as well as depth search strategies are combined to optimize system responsiveness and solution accuracy.

6.2.3 The Inference Engine

The other part of an expert system is the inference engine. The inference engine is where reasoning occurs as the AI program is executed. The inference engine is an integral part of the language being used to write the system. There are a number of ways for the reasoning to occur depending on the way knowledge has been represented. Two of these are described below.

Rule-Based Reasoning

When the knowledge in the system is rule-based the reasoning is accomplished by chaining the rules. Rule-based reasoning is powerful for the analysis of problems in complicated domains. Rules allow users to describe the heuristic and procedural knowledge of a domain. Reasoning can also be accomplished from first principles. This means the basic physics or laws of the domain are used rather than heuristic knowledge.

Frame-based Representation

When the knowledge is frame-based the reasoning is accomplished by matching various frames. Frames are patterns for describing and recognizing repeating sets of features. Frames allow the users to describe objects of the user's domain. The attributes of the objects can be declarative or procedural and can be inherited from other objects using semantically precise rules.

Other approaches to reasoning are the use of probability, causal-dependency, formal logic, analogy, generation and test, marker propagation, fuzzy reasoning, and procedural arrangement.

6.3 COMMERCIAL SYSTEMS

There are a number of computer systems, usually called workstations, which are used as tools for implementing software programs and particularly AI-based programs. These tools have evolved to meet the needs of specific higher level languages. The tools include many automatic functions to make programming faster and less error prone. They also use different machine architectures which are more appropriate to the type of language used. Examples of specialized computers are the LISP machines for implementing AI systems and the Japanese fifth generation computer which uses PROLOG.

Commercial AI system tools can be categorized as follows:

1. General AI languages. An example of this is LISP. These can be supported by many of the currently popular conventional computer operating systems and workstations even to the level of the PC.

2. Special purpose AI languages. Examples are OPS and PROLOG. These also are supported by most operating systems.

3. Data base manipulators. These include EXPERT-EASE,
    K-BASE, and RULE MASTER. They provide support for
    interfacing AI programs and data bases.

4. Prototype construction tools for building expert systems
    like M1, Texas Instruments PERSONAL CONSULTANT
    which runs on their large PC, and Tektronix systems.

5. Master carpenter tool kits for building and running major
    systems. This list includes ART, KEE, LOOPS, S1, SRL
    LISP machines, and many others. They often run on
    workstations like APOLLO, SUN, DAISY, MICROVAX
    and the Xerox 1100 series.

Today the users of these tools must know the science of AI
and computers. In the future this may no longer be required.
Some of these systems are evolving to a state where they will become so specialized that the user no longer needs to know computer science to use them. A simple example of this trend is the
very effective use of spread sheets by PC users who know relatively little about computers. A more sophisticated example is the
system, PLANPOWER, produced by Applied Expert Systems.
This system does individual financial planning by combining the
skills of many experts in a way that allows them to service their
customers much more rapidly. In this case the incentive was the
advent of financial deregulation. To take advantage of this new
market, information technology was adopted as a strategic business weapon by the user. The result was an expert system tailored
to the needs of this particular specialized business.

6.4 IMPLEMENTATION STRATEGIES

As in any other form of FIS, guidance from and to the user is required for the implementation of the parts that use Artificial
Intelligence programs. The guidance is derived from a determination of better manufacturing procedures and controls made possible by the use of AI technology. The improved procedures and
controls have to be defined jointly by the users and implementors
using the techniques discussed in Chapters 2 and 3. Also, everything about system implementation discussed in Chapter 8 continues to apply. This section deals with unique considerations,

resulting from the nature of AI, that can assist the user in understanding how to proceed with the AI portions of a manufacturing decision support and control system.

A minimum of two disciplines are required to design the AI programs. These disciplines are usually represented by (1) an expert from the factory who is sometimes called the domain expert and (2) a knowledge engineer with extensive AI experience. The domain expert knows in detail the problems confronting the particular part or domain of the factory where AI programs can potentially be of assistance and what information is available. The knowledge engineer knows what and how AI technology can be applied to general problem solving. This individual also knows how but not what information to collect from the domain expert. Together they can formulate the problems in terms that are useful and appropriate for implementation.

This team identifies problems and develops concepts for their solution. These are then formalized in a way that allows others to help with the implementation tasks. Techniques for accomplishing this are discussed by Waterman et al. (1983). There are some essential questions that the team will ask in order to identify and formulate the manufacturing problems. Some examples follow:

1. What knowledge about the manufacturing process or problem is given and what is inferred?

2. What are the strategies and hypotheses used by manufacturing management to guide the operation?

3. How are the objects such as machines, tools, people, products, processes recipes, and so forth related?

4. What is the hierarchy of the objects relationship and management relationship?

5. What are the processes currently used to solve existing problems and what are the constraints applied to these processes during problem solving?

6. What is the information flow?

7. Can knowledge about how to solve problems be separated from the justification of their solution?

Implementation is carried out by the very early construction of prototype programs and systems. This can be done because the writing of systems using AI languages is much faster and more efficient than with conventional languages. Implementation then becomes a continuing series of revisions to the existing prototype. These revisions can be changes in the AI programs, changes to different computing machines and operating systems, conversion of some of the AI code to conventional code to speed up portions of program execution, and the addition of new programs to accomplish new tasks.

System implementation through prototype revision is very helpful to the users because it gives them a chance to actually try out the emerging system and participate in its evolution. As stated earlier, successful systems will continue this evolutionary process for the life of the factory. Thus the system itself should never become obsolete!

6.5 LEGAL ASPECTS OF AI

The legal community has not yet defined rules governing liability from the results of AI program use. Authors of books or articles covering the same topics as an AI program are not liable for incorrect information that leads to losses when this information is applied. It is not clear that the author of an AI program which produces an incorrect diagnosis or activates the wrong control enjoys similar protection. This has significantly slowed the application of expert systems in fields such as medical diagnosis and may well curtail the development and sale of commercial turn-key manufacturing decision support systems based on AI technology.

6.6 GLOSSARY OF AI JARGON

Assertion: A declared statement of fact.
Backtrack: A way of searching where each descendent of a hierarchical tree is evaluated in depth-first fashion until a solution or a dead end is found. If a dead end is found the search backs up and tries another path of the most recent branching.
Backward search: The search begins with the goal to be proven and attempts to satisfy all the conditions that would make the goal true. This involves searching for rules that have the

goal as a conclusion and continuing back until the start states are found. (This is called backward chaining or goal-driven.)

Breadth-first search: A method of searching a hierarchical tree where every immediate descendent at the same level is evaluated before going to the lower levels of descendents.

Decision tree: A representation of paths with branch points at nodes in the tree. The paths represent logical alternatives when conditions are applied by if-then-else statements at each node.

Depth-first search: The search of a tree when the deepest left or right most descendent is evaluated before the siblings.

Domain: An activity or area of expertise in which problems are being addressed by an AI system.

Forward search: A search that begins with everything known about a problem and attempts to draw conclusions based on the data. (This is called data-driven.)

Heuristics: Rules of thumb based on experience used to reduce the size of the search space in an AI problem.

Inference: A logical conclusion reached by the art of reasoning.

Inference engine: The part of an expert system that reasons about and controls the knowledge base. It is usually in the form of production rules.

Inheritance: Every object in a class inherits properties, routines, values, and so forth from the class it belongs to.

Instance: An example or occurrence.

Knowledge acquisition: The collection of knowledge and its conversion to a machine usable form.

Knowledge base: The part of an expert system containing domain specific knowledge.

Knowledge engineer: A person trained in knowledge representation and search techniques who works with a domain expert to develop an expert system.

LIPS: A measure of the speed of a processor used to make logical inferences in logical inferences per second. A LIP is about equal to 50 machine instructions.

List: A data structure.

Logical connectives: Symbols that define the relationships of "AND", "OR", "IMPLIES", and "NOT".

MIPS: Millions of machine instructions per second that measure computing maching performance.

Node: Each decision point in a search tree.

Object: A data structuring technique in which objects have certain properties including inheritance, attributes (with values), and embedded procedures. Parameter values passed to these procedures by the program generate new instances of the object.

Object-oriented programming: A highly modular programming style that associates descriptive and procedural attributes directly with objects. Since each object has its own procedural characteristics it can perform local actions such as display or modify itself and it can both receive information from and return information to other objects.

Opinion Based System: A Knowledge Based system.

Paradigm: An example or model.

Penetrance: The minimum number of nodes in a search tree required to get the solution divided by the total number of nodes traversed.

Predicate calculus: A system used in mathematics for representing and proving logical relations.

Production rules: Rules on how to process knowledge to reach a conclusion.

Recursion: Repeating by going back to some preceding part of a program or calculation followed by a new program path or calculation until some condition terminates the action.

Semantics: The general theory of interpretation and symbolic meaning. This includes the characteristics of symbols and their relationships, i.e., what an expression means. Frames and scripts are used to describe semantic meanings.

Space: A set of possible states.

State: A representation of a "situation."

Syntax: The way in which words are put together to form an expression. A measure of how well-formed an expression is.

Unification: A technique used to match patterns of terms and variables in assertions and propositions to supply answers to queries and identify variables.

REFERENCES

Gustaso, W., and Ulrich, P.E. 1973. *Elementary Symbolic Logic.* Holt, Rinehart and Winston, Inc., N.Y.

Hadavi, K. November 1985. "Dynamic Scheduling for FMS", Conference Proceedings of Autofact '85, Computer and Automation Systems Association of the SME. Dearborn, Michigan, page 6-59.

Haugeland, J. 1985. *Artificial Intelligence: The Very Idea*. The MIT Press, Cambridge, Massachusetts, page 23.

Waterman, D., Hayes-Roth, F., and Lenat, D. 1983. *Building Expert Systems*. Addison-Wesley Publishing Co.

ADDITIONAL REFERENCES

Ferguson, R. November 1981. "PROLOG—A Step Toward the Ultimate Language." BYTE Publications Inc., page 384.

Gallagher R. T. Oct. 6, 1983. "Logic—Programming Era Is Dawning." Electronics, page 110.

Nilsson, N. 1980. *Principles of Artificial Intelligence*. Tioga Publishing Co.

7

Communication Networks
for Factories

7.1 INTRODUCTION

This chapter describes the kinds of data communications that are
applied to meet the needs of manufacturing. Emphasis is placed on
the local area network (LAN) as the means for transporting information and the manufacturing automation protocol (MAP) as the
standard for its transmission. The more general concepts of distributed computing within a distributed computer system are beyond the scope of this chapter and, therefore, are omitted. Distributed computing has a distinct potential advantage as was explained in Chapter 1 and by Davies, et al. (1981). It is, however,
a consideration separate from communications.

The communication network is the backbone of a factory
because it is the conduit, connected to all parts of the factory,
which supplies information and control directives. This common

network permits any production operation to communicate with any other operation or function of the factory. It therefore makes possible the level of communication required to support the basic factory needs which were reviewed in Chapter 2 and the product tracking and flow control requirements of flexible manufacturing systems described in Chapter 4.

Local area network technology is appropriate for factories because it supports modularity. Work cell controllers, product reporting points, and higher level data analysis computers can be joined to or removed from the LAN without affecting communication to other cells. This provides a means for the cost effective, evolutionary modification of the factory configuration and equipment as changes occur in markets, products and production process technology. The distributed FIS system described in Chapter 1, Figure 1.3, makes effective use of the LAN for communications.

The need for a communication system appropriate to factories has evolved historically. During the past 25 years a form of clustered factory automation occurred as groups of closely related production equipment were brought under the local control of small computers. These equipment clusters are often referred to as "Islands of Automation." An island of automation is composed of the work cell or cells where the manufacturing process is logically self-contained and therefore naturally subject to local control. The control of these islands usually resides within the cells' processing equipment which has been automated by the equipment vendor to achieve better yield and product quality. Communication within the island is therefore a part of the processing equipment and usually customized by each vendor for each application. Many vendors of this equipment provide a communication port for access to systems outside of the island. The port usually supports the RS232-C communication standard described later in this chapter. Unfortunately this port alone is often not sufficient to support the required communications because other software and parameter definitions are missing.

The task of the FIS communication network is to provide information exchange between the islands of automation. The connection of an island or a cell within the island to an external communication network, after the cell has been designed as an independent entity, requires custom hardware and software to deal with each particular type of controller, data type transferred, and response expected. This has been a labor intensive task because there have been no communication network standards to guide the design of these cells which form the island. In the absence of a standard, each island has been designed to perform a specific processing operation independent of any outside influence. There is now, however, a strong incentive for inter-island communication as the benefits of manufacturing system integration became apparent.

This incentive resulted in large manufacturing companies like General Motors selecting in the early 1980s a single manufacturing automation protocol (MAP) for the factory. This has greatly simplified the communication interface by providing a standard to which processing equipment vendors, computer equipment designers, and users of manufacturing systems can now subscribe.

The following sections will review communication networks, the standards used for communications, and explain some basic terminology. From this the reader may gain a better understanding of factory communication systems in general and the significance of the recent movement toward international agreement on communication standards for manufacturing.

7.2 MODERN COMMUNICATION NETWORKS

7.2.1 Network Topologies

There are six possible topologies associated with communication networks. These are shown in Figure 7.1. The three most popular for factory applications are the star, ring, and bus. The star is the classical hierarchical system with a single computer handling the communications to the rest of the system. This configuration has the disadvantage of having a single point, the central computer, which can fail and totally disable the system. The ring passes the messages from node to node, thus providing a way to amplify signals and therefore a longer distance capability with somewhat slower response time. By physically bypassing inoperative nodes and using bidirectional transmission, the ring can overcome the disadvantage or reduced overall system reliability. This is the product of the reliability of each node between the sender and the receiver. The bus topology does not use regeneration and therefore

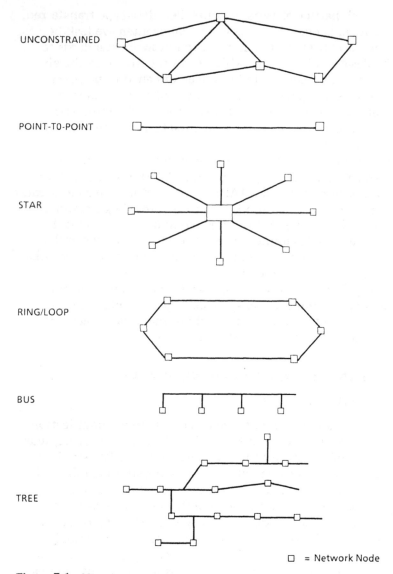

UNCONSTRAINED

POINT-TO-POINT

STAR

RING/LOOP

BUS

TREE

□ = Network Node

Figure 7.1   Network topologies.

does not have the distance capability of a ring. It is immune to failure from any specific node and therefore has an inherent advantage for short run factory applications.

7.2.2 Network Evolution

In the past few years communication technology appropriate to factories has enjoyed a rapid evolution made economically feasible by the development of low cost digital integrated circuits. The first local net conference was held in 1976. Ethernet, a local area communication system developed by Xerox, was introduced in 1979. It became the accepted approach to communications for office automation. The manufacturing automation protocol (MAP) token passing system was proposed in 1982 under the sponsorship of General Motors. This is described later in the chapter.

The initial networks for telephones used circuit switched continuous transmission for point-to-point communications. This type of network topology is shown in Figure 7.1. In this system the routing of messages is preallocated so there is only one route for a message to travel from one point to another in the network.

Now public branch exchanges (PBX) and computerized branch exchanges (CBX) which use unconstrained networks shown in Figure 7.1 are the mainstays of the telephone communications business. This industry has integrated data transmission and switching systems with these exchanges. These are worth consideration for some types of office automation but are not appropriate for factories.

The inefficiencies of continuous transmission were overcome by the development of the value added network (VAN). The transmission changed from continuous to a form called packet switched. A packet is a series of sentences or other forms of data grouped into a bundle or packet. The packet also contains information pertaining to its source and destination. The communication system is thus able to mix packets and dynamically route them to their destinations. Packet switching permits a much larger and more diverse telephone system to be realized based upon the use of satellite, optical fiber, microwave, and other means for transmission.

Fortunately, the local area network (LAN) next evolved from the extensive development effort applied to office automation.

LANs are ideal to support factory communication providing they are reliable and have predictable transmission characteristics. Most office automation LANs use a form of transmission and protocol initially developed by the XEROX corporation called Ethernet. This is very appropriate for offices because system transmission time degradation under heavy load conditions is acceptable in an office environment. It is much less acceptable in many industrial applications where lack of timely message service can result in loss of process control. Unfortunately the design of Ethernet does not support the predictable response times required for manufacturing applications.

The use of Token Passing LANs described in the next section has solved the problem of performance predictability. That is why they are being adopted by the worldwide manufacturing community as the communication technology for factories. A compatible standard, similar to Ethernet, called TOPS is also being introduced to handle office communications in factory environments.

7.2.3 Local Area Networks

A local area network (LAN) is a localized data communication system employing a shared communication medium, packetized data transmission, and highly distributed control with no single point of failure. A LAN is characterized by a number of attributes. It is usually owned by a single organization over a limited geographical area of approximately 8 miles. (There are some LAN applications, however, extending over 100 miles.) The bandwidth is usually 10 megabits per second minimum using a relatively cheap coaxial or fiber optic cable. Connections to the cables are done with small hardware devices called alternatively the cable interface unit (CIU), network interface unit (NIU), communications server, terminal server, or bus interface unit. The interface is designed to support the entire spectrum of hardware and software conditions and is thus termed universal. It buffers and manages the flow of data packets into and out of the LAN.

The primary use of LANs is for the transmission of data rather than voice or video. The transmission is via packet switching with the resulting advantages of address control.

There are two basic ways to access a LAN: permission and contention.

Permission Access

Access by permission is implemented by using a token which is passed from station to station or node to node. If a node wants to transmit a message packet it waits until the token arrives and then transmits. When the transmitting station has sent a packet the token is passed to an adjacent node. The token passes from node to node in a preprogrammed manner, residing at each node only long enough to determine if a packet is waiting to be sent. With the ring topology shown in Figure 7.1, the token is passed to the next physical node in the ring.

A token bus has the advantage of including in the token an address to which it is to go next. The passage of the token can thus be controlled by a logical set of nodes superimposed on the physical bus. This permits the token to be sent more frequently to nodes that require a higher priority of transmission. This type of access control is also very predictable under conditions when the communication traffic is dense. Under very heavy loads, all nodes get periodic access within a predictable time-frame. These are the reasons why the token passing bus has been selected for use in factories where peak data transfer loads have to be executed and the priority of process communications are essential to maintain manufacturing control.

Contention Access

Most contention accessed LANs work on the principle of carrier sense multiple access with collision detection (CSMA/CD). This simply means that the transmitter and receiver are both active at any time to provide the kind of performance described below. Before a node will initiate a transmission it will listen and determine if any other node is transmitting. If a carrier is detected (sensed), the node that wants to transmit will wait until the transmission is complete. When no carrier is detected the node will begin to transmit. At this time it may happen that other nodes have also been waiting to transmit packets. This is why the system is called multiple access. Many nodes could then try to transmit at approximately the same time. Because of the transmission delay as the electromagnetic wave travels along the cable, each node thinks at the beginning of transmission that no other node is transmitting. During transmission the sending node continues to listen for any signal other than the one being sent. If a carrier from another node is sensed by the receiver, the assumption is that a collision in communication will occur because of simultaneous transmission from the multiple nodes. The node that first detects a signal from another node stops transmitting for a period of time. The period is at least as long as it takes the electromagnetic wave of the carrier to travel twice the length of the LAN cable for the site in question. The second node will also detect the signal and stop transmitting. Each sending node will wait a different programmable time before trying to transmit again. All messages that potentially collide are canceled before the nodes again try to send their messages. When many users try to simultaneously use the system, contention access can result in unpredictable message transmission time because of this thrashing for resource allocation. The CASMA/CD technique is particularly effective with a large number of nodes where access time is not critical. It gives excellent response for bursty data traffic on lightly used systems. These characteristics led to its selection for office system applications.

7.2.4 LAN Comparisons

The characteristics of bandwidth, access method, and topology are compared for local area networks in this section to gain insight as to why the particular configuration for the MAP system discussed in the next section was chosen.

Bandwidth

Bandwidth describes the size of the spectrum of frequencies available for message transmission. Base band or carrier band has a narrow bandwidth and is the lowest cost, is easiest to install, and runs at the highest speed. It unfortunately must have active repeaters for long distance and suffers from reflected signals, especially when the stations are closely packed.

Broadband offers multiple channels that can be matched to various applications including channels for video and voice as well as data. A full tree topology can be implemented with low cost passive CATV cable. Extra cost and complexity comes from the need for rf modems at the stations. These are subject to frequency drift which can result in loss of transmission. A more expensive headend than that required for the carrier band system is necessary for line termination.

Access Method

CSMA/CD is simple, effective, low cost, and supports a large number of stations. Performance is limited by the product of speed and distance. The circuits for collision detection are sensitive and require high quality design and components to be reliable. Although 10 megabits per second (mbps) is the commonly used transmission speed, the stochastic nature of collisions makes the effective speed more like 3 mbps. Under heavy load the CSMA/CD method of access causes excessive and unpredictable delays in transmission time. The approximate packet transmission time increases by an order of magnitude, (10 times) at 60% transmission load where the token access does not degrade to this level until about 85% load.

Token access is collision free. Under light traffic loads it is slower than CSMA/CD but under heavy loads it is considerably faster and predictable. It is speed/distance insensitive with about the same effective transmission rate as CSMA/CD. The negative aspects are the more complex configuration (if ring is used) and the management of network restart if the token is lost or when node additions are made. Failure of a repeater can also shutdown sections of a channel.

Topology

Ring topology, shown in Figure 7.1, offers closed circuit integrity. This means if the loop is severed the transmission can be achieved by traversing in the other direction. The inherent point-to-point transmission allows many different types of media, including fiber optics, to be used in the same system. Because the active stations are repeaters, the distance traveled can be very long but of course the repeaters can fail.

In contrast to the ring, branching bus topology allows low cost station taps which can be added without interfering with network functionality. Standard amplifiers are used but the bus is not suitable for fiber optics because of the difficulty of making the line tap connections.

7.3 MANUFACTURING AUTOMATION
   PROTOCOL (MAP)

7.3.1 History

In 1979, the Engineering and Manufacturing Computer Coordina-
tion activity of the General Motors Corporation sponsored a Local
Communication Networks Users Group to facilitate information
exchange pertaining to plant-floor computer data communications.
On the recommendation of this group, a task force was formed to
identify and evaluate existing communication techniques with the
objective of defining a common standard for plant-floor systems.
The task force was chartered by the Corporate Computers In
Manufacturing Steering Committee in November, 1980, as the
Manufacturing Automation Protocol Task Force (MAP).

The goal of MAP was to prepare a specification that would
allow common communication among diverse intelligent devices in
a cost effective and consistent manner. To accomplish this, MAP
established a set of three objectives.

1. Define a MAP message standard which supports applica-
    tion-to-application communication

2. Identify application functions to be supported by the
    message format standard

3. Recommend protocol(s) that meet these functional re-
    quirements

The first MAP document, published in October, 1982, provided guidance for general network considerations and implementation information. Numerous revisions and additions to this document have occurred in the following years as the proposed standards have been refined and accepted by the international manufacturing community.

The standard is based on existing and proposed standards for
local area networks. It is therefore not a new protocol but a set of
protocols chosen from existing documented and implemented
procedures which will continue to evolve as the technology available to support communications changes. The intent is to meet the
need to interconnect the many pieces of manufacturing equipment,

supplied by different vendors, with a standard communication network that is consistent with a factory environment. The MAP standard thus promotes a multi-vendor manufacturing system. The MAP architecture is based on the International Standards Organization (ISO) reference model for open systems interconnection (OSI).

7.3.2   The Open Systems Interconnection (OSI) Model

This model is supported by most national and international standards organizations for communications. Basing MAP on this model insures a measure of compatibility between international vendors of manufacturing and communication equipment. Specific modules or layers of the model are tailored to meet the needs of manufacturing applications. This results in MAP conforming as a subset of the OSI model.

The OSI model presumes a modularization of the networking support software based on functionality. Each module is in the form of a layer in the model. Each layer provides selected network services to the layer above. The services are implemented through programs in the layer and through services available from the layers below. In theory, any layer can be replaced by a new layer which provides the same services in a different way without affecting the user's perception of the network operation. This allows different vendors of communication software and hardware to compete using different technologies for the markets defined by each layer. Networks can in this way be configured to meet specific needs while using common, competitively priced, components.

The lowest layer is the physical media over which data is transmitted. The highest layer supports the application programs that use these data. Multiple sublayers can be used to implement a given layer, or layers may be omitted if the functions are not required. The absolute requirement is that the standards must be adhered to at each interface between layers. These layers are shown in Figure 7.2. Some of the organizations working on various parts of this specification are shown in Figure 7.3.

The following is a summary description of the MAP functions and specifications currently appropriate for each layer.

LAYERS	FUNCTION
USER PROGRAM	APPLICATION PROGRAMS (NOT PART OF THE OSI MODEL)
LAYER 7 APPLICATION	PROVIDES ALL SERVICES DIRECTLY COMPREHENSIBLE TO APPLICATION PROGRAMS
LAYER 6 PRESENTATION	RESTRUCTURES DATA TO/FROM STANDARDIZED FORMAT USED WITHIN THE NETWORK
LAYER 5 SESSION	NAME/ADDRESS TRANSLATION, ACCESS SECURITY, AND SYNCHRONIZE & MANAGE DATA
LAYER 4 TRANSPORT	PROVIDES TRANSPARENT, RELIABLE DATA TRANSFER FROM END NODE TO END NODE
LAYER 3 NETWORK	PERFORMS MESSAGE ROUTING FOR DATA TRANSFER BETWEEN NONADJACENT NODES
LAYER 2 DATA LINK	IMPROVES ERROR RATE FOR MESSAGES MOVED BETWEEN ADJACENT NODES
LAYER 1 PHYSICAL	ENCODES AND PHYSICALLY TRANSFERS MESSAGES BETWEEN ADJACENT NODES

Figure 7.2   MAP protocol layers.

Layer 1, (Physical)—Encodes and physically transfers messages between adjacent nodes of the network. Uses IEEE 802.4 token access on a broadband media.

Layer 2, (Data Link)—Improves error rate for messages moved between adjacent nodes. Along with layer 1, allows for the interconnection of nodes from different vendors. Protocols at layer 1 and 2 must be consistent with adjacent connected network nodes. Uses IEEE 802.2 link level control.

Layer 3, (Network)—Performs message routing for data transfer between non-adjacent nodes. ISO/NBS Internet standard is under consideration.

Layer 4, (Transport)—Provides transparent, reliable data transfer from end node to end node. Uses ISO/NBS Transport class 4.

Participant / Layer	Nestar	3 COM	IBM	DEC	SYTEC		MAP	
7 Application	Nestar	3 COM	SNA	DNA	SYTEC		Future & Existing Standards	
6 Presentation	Nestar	3 COM	SNA	DNA	SYTEC		Null	
5 Session	Xerox	Xerox	SNA	DNA	TCP/IP		ISO Minimal Session Kernel	
4 Transport	Xerox	Xerox	SNA	DNA	TCP/IP		ISO Transport Class 4	
3 Network	Xerox	Xerox	SNA	DNA	TCP/IP		ISO Connectionless Internet	
2 Data Link	Arc Net	Ether Net	Token Ring	Ether Net	HDLC	IEEE 802.2	IEEE 802.2	IEEE 802.2
1 Physical	Arc Net	Ether Net	Token Ring	Ether Net	SYTEC	CSMA/CD 802.3 Boeing	Token Bus 802.4 GM	Token Ring 802.5 IBM

Figure 7.3   Network integration paticipants.

Layer 5, (Session)—Provides name and address translation, access security, synchronization and data management. NBS Session is under consideration.

Layer 6, (Presentation)—Restructures the data to/from the standardized format used within the network. Existing standards are under consideration.

Layer 7, (Application)—Provides all the services directly comprehensible to the application programs. Existing standards are under consideration.

User Program—Specific application programs are not part of the OSI model. These programs will not conform to any standard and are specific for each user task.

7.4 NETWORK DESIGN AND PLANNING

The network should be planned in a series of steps to ensure that
it is compatible with the needs of the manufacturing site or sites.
One approach is to first define the objectives in terms of price-to-
performance, expected life time, and degree of flexibility. These
should support the business objectives. If there are no business objectives which require the factory information system and therefore the communication network, there is no basis to continue the
project.

Once the objectives are established there is a significant
amount of work needed to collect data related to the manufacturing environment, equipment, communication traffic, requirements for communication, information flow, and so forth. Much
of this is naturally included as a part of the needs analysis discussed in Chapters 2 and 3, and specific functions that are defined
as a result of considerations reviewed in Chapters 4 through 6.
This work should culminate in a document that describes the applications, costs, benefits, design criteria, and conceptual design.
This document leads to the system requirements specification.

The requirements specification is used as a guide in the selection of a network vendor. Alternatives, better cost estimates, performance evaluations, and technology trade-offs are considered together with potential vendors. The result of this effort is a specific
recommendation for a communication system and supplier. The
implementation of the system by the selected vendor and the required staffing to maintain and operate the network can then occur. It is not feasible for most manufacturing organizations to
build and install their own LAN network. They can, however,
learn to maintain the system as described in the next section.

7.5 MAINTENANCE OF COMMUNICATION NETWORKS

Local area networks are more difficult to understand and maintain than simple point-to-point communication systems. Most
small factories do not have and can not attract or support the type
of staff required for this maintenance. It is therefore best to start
with a system supplied by a reliable vendor who will support it
through a maintenance agreement for the first few years of use.

As the use of computers to control and manage the factory increases, an ever increasing need develops to have competent computer systems engineers and software engineers at the manufacturing sites. The number and capability of these people will grow until there will be an appropriate time when they can take over the maintenance task. Until then, the selection of the network vendor is critical.

7.6 COMMUNICATION BASICS

This section is included to provide an understanding of fundamental communication technology and nomenclature.

7.6.1 Representing Information

The objective of communication is the transfer of information. To do this, the methods used to represent the information must be defined.

Bits

The fundamental representative of information in a digital computer is the bit. It represents the state of a digital circuit element. These binary circuit elements can be in only one of two states, on or off. The voltage appearing at the input and output of the circuit element depends on the state and is referred to as high or low. These voltage levels are usually represented by a 1 or a 0.

Bytes

To represent characters, the bits are combined to represent more information choices than the two on-off states of a single circuit element. For example, a pair of bits can represent four unique characters as shown in Table 7.1.

It takes 7 bits, (128 combinations of 1s or 0s) to reasonably represent the English alphabet, numbers, and the control characters used for communication. This combination of bits plus one more for the validation of data transmission is called a byte. The byte can also refer to other bit configurations but the 8 bit byte is the one usually used for factory communications.

Table 7.1 Combining Bits

Character	First Bit	Second Bit
1	0	0
2	0	1
3	1	0
4	1	1

ASCII

The pattern of bits represented by a byte has to be translated into a recognizable character. The most common translation code used for general data transmission is the American Standard Code for Information Interchange (ASCII). This is verbally referred to as "asky". It uses 7 bits for defining 128 different characters and the most significant 8th bit to test for transmission errors. Error detection is usually accomplished by using a combination of vertical and horizontal parity checking described by Deasington (1982). The capability of ASCII to efficiently validate transmitted data and its extensive hardware and software support supplied by the computing industry makes it today's preferred code for general manufacturing communications.

Baud Rate

The term baud is used to define the rate of information transmission in terms of signal rate per second. If we limit the case to the usual binary bits described earlier with only two states, 0 and 1, then baud is equivalent to bits-per-second. If the case is further limited to using ASCII code where 10 or 11 bits define a character, we can easily convert bauds to characters-per-second by dividing baud rate by 10 or 11 bits per character.

Physical limits to the rate at which bits can be transmitted depend on the length and characteristics of the transmission line or channel. Twisted pairs of shielded wire are capable of rates up to 9600 bauds and have been used above 10 megabits per second in cases where the quality of the cable is good and the lengths are relatively short. Broader bandwidth coaxial cable or optical fiber

cables can transmit at many millions of bits-per-second. For today's factory applications involving data from floor operations,
9600 bauds is perfectly adequate as long as video displays are not
required.

RS232-C

To specify a communication system, a number of additional
parameters have to be defined in addiction to a code like ASCII to
convert bits to characters. These are, (1) what signals need to be
provided to control the communication equipment, (2) at what
voltage level these signals are generated, and (3) by which party,
the sender or receiver. In the United States the Electronic Industries Association (EIA) provides the standards that define these
parameters. The standard most frequently used for manufacturing
is the EIA RS232-C. The minimum usable set of this standard consists of the eight functions listed below.

1. Protective ground

2. Reference ground

3. Transmitted data

4. Received data

5. Request to send

6. Clear to send

7. Data set ready

8. Carrier detect

A sense for communication command exchange is gained
from these basic functions. The reader can again see Deasington
(1982) for further details which go beyond the scope of this chapter.

7.6.2 Types of Communications

Once information has been adequately represented it is available
for transmission. This section describes some of the transmission
techniques.

Serial and Parallel Transmission

The two voltage levels for each of the 8 bits that make an ASCII byte can be transmitted in parallel through eight channels or in series through one channel. Parallel transmission is expensive for the long distances required in most factories because of the cost of the physical cable with eight conductors. Parallel transmission is therefore used only for short connections between computers or computers and peripheral equipment such as printers. Its advantage is the ability to transfer data at a higher rate than serial transmission.

Serial transmission through a single channel requires the sender and receiver to define and recognize the beginning and end of each byte. The stream of bits flowing through the channel can then be accumulated by a shift register until a byte is received. When this is accomplished the byte is then transferred in parallel to appropriate buffers.

Synchronous and Asynchronus Transmission

There are two methods commonly used to determine the beginning and end of each byte. The two methods are called synchronous and asynchronous transmission. Synchronous transmission requires the receiver to know the time when a byte will arrive. The timing is provided by periodically synchronizing transmission with a clock signal from some control source.

When transmitting asynchronously the byte can be sent and received at any time. A start bit is generated by the sender at the beginning of each byte followed by the 8 data bits and one or two stop bits. This bit stream is shown in Figure 7.4. A clock at the receiver is started by the start bit and runs for eight ticks to define the arrival time of each of the eight data bits. Asynchronous transmission is used for most computer system peripherals because it is simple and the clock is self-contained. It is therefore very appropriate for factory applications where local input-output devices such as terminals and printers are widely scattered throughout a plant.

Narrow and Broadband Transmission

The transmission of the square wave response shown in Figure 7.4 requires a transmission media that is capable of operating at fre-

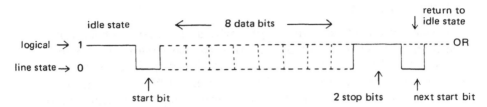

Figure 7.4   The byte-bit stream.

quencies up to 10 megahertz. A frequency bandwidth of about 10 megahertz is necessary to make the corners of the wave-form approximately square instead of rounded. Ten megahertz is also close to the maximum frequency supported by today's lower cost integrated circuits. Networks that operate at this frequency are called base band or carrier band systems. Their bandwidth, or range of frequency response, can support the transmission of only a single channel but they are simple compared to broadband systems and fulfill the majority of today's needs for factory information systems.

If the transmission media, transmitters, and receivers are capable of higher frequency response a broadband form of communication can be used. In this case the available spectrum can be divided into 10 megahertz bands and each of these can be used as a separate channel. This added capability and flexibility for additional data (and even video) channels is an early consideration when selecting the type of cable for a factory installation.

Simplex and Duplex

A channel which allows data to travel in both directions is called duplex. If the channel can support data transmission in both directions simultaneously it is called full duplex. Channels that can support data flow in both directions but only one direction at a time are called half duplex. If the channel can only transmit in one direction, such as a television channel, it is called a simplex channel. Most factory systems are full duplex so that bilateral communications for error correction and alarms can be executed in a timely manner.

Packet Switching

A packet is a finite stream of bytes with an identifiable beginning
and end. These bursts of information can be time multiplexed
as described later in this section to provide very flexible and far
reaching communications. Packet switching fulfills the need of
many manufacturing operations to communicate with other remote
locations such as suppliers, subassembly operations, and perhaps
the head quarters of the division which may even be in another
country. This is usually accomplished by using existing telephone
lines.

The packet, or group of bytes, is collected and assembled by
a computer module called a packet assembler-disassembler (PAD).
The packet is usually composed of from 60 to 1000 characters.
The header, or characters at the beginning of the packet contain
the following information:

1. Preamble—indicates a message is coming. This can be
   specific to the network used for transmission.

2. Destination address—defines where the packet is to go.

3. Source address—defines the source of the packet.

4. Control—asks for acknowledgment without having to
   read the data.

This header is followed by the data stream. At the end of the
packet a frequency check sum (FCS) of 16 to 32 bits confirms all
the packet bits including the preamble, destination, source, and
control bits. Any errors detected by the FCS are used to alter control of message flow (such as "repeat the packet") and are stored
for later evaluation by those responsible for the packet switching
network.

X.25

The International Telegraph and Telephone Consultative Committee (CCITT) has defined standards for packet switching because it
is used extensively for international communication. The specific
standard that is most favored today is called X.25. X.25 defines
three levels called the physical, link, and packet. The physical level
is similar to RS232-C described earlier in this chapter. RS232-C is,

therefore, often used to fulfill this function. The link level defines the protocol for error free, transparent packet transmission. This protocol is called high level data link control (HDLC). It is an old standard that has been used to meet this need. The packet level defines the structure of the packet itself.

An important concept here is the modularity of each of these levels. The modularity is achieved by defining each level's interface to the adjacent level at the detail required for controlled interlevel communication. As long as each level provides the correct interface what happens within the level does not affect system capability. This means each level can be independent, designed and produced by different vendors, and it can use different design approaches. For example, the packet level specification defines 15 groups of logical channels of communication for each physical channel that exists in the telephone network. Each of these groups are in turn divided into 255 logical channels. The groups are defined by a 4 bit code and the logical channels by an 8 bit code. All of this information in the header of the packet allows the link level above the packet level to control packet transmission. How the packet gets assembled and disassembled is not important to the link level as long as the results are compatible with the interface specification used by the link level to accomplish its control. This modular independency becomes even more important as we move to higher communication levels such as those previously described for local area networks (LANs).

7.6.3 Concentrators and Multiplexors

The system used to transfer information also must be cost effective. This section describes two types of hardware that can assist in reducing costs.

There are many instances in a factory where a number of terminals, testers, and pieces of processing equipment need to communicate to a computer or to a number of other devices. It is often economical to make this connection through a single high-speed data line instead of separate lines connected between each of these devices. This approach is particularly applicable when the devices are clustered together as is often the case in work cells. To accomplish this the high-speed data line is divided into communication subchannels, one for each device. The formation of subchannels is done either by a concentrator or a multiplexor (MUX).

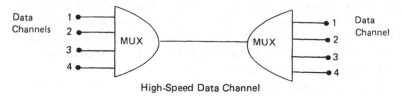

Figure 7.5   The multiplexor function.

Concentrators

The concentrator connects many devices on the factory floor to a
single computer. There is therefore only one concentrator per application. It is controlled by software programs that allocate the
subchannels and convert the character-by-character transmission
from the devices into blocks of data transmitted to and from the
computer. The operation of the concentrator is completely transparent to the user. This means the manipulation of the data
streams to accomplish transmission is automatic and not dependent on the source data being transmitted.

Multiplexors

The multiplexor connects one or more devices to one or more
other devices by means of a data channel as illustrated in Figure
7 .5. Multiplexors at each end of a high-speed data channel provide
a number of subchannels. MUXs are therefore always used in
matching pairs. At one end of the high-speed data channel, a different device can be connected to each subchannel of the MUX.
At the other end, a second MUX provides these same channels to
other devices. The MUX uses hardware or firmware to do channel
allocation and transmission control.

Multiplexing can be done in two ways, by frequency division
or by time division. Frequency division multiplexing divides the
available frequency spectrum into frequency bands as was briefly
discussed under the preceding topic of bandwidth. Each band is
used as a channel for communication. Transmitters and receivers
are tuned to each band thereby achieving isolation from other
bands.

Time division multiplexing establishes a separate time period
during which each channel will transmit and receive. Some forms

of demand multiplexors can dynamically allocate the length and frequency of these transmission periods depending on the traffic load. This type of multiplexing is very similar to packet switching systems except the source and destination of the transmission is determined by the physical channel connections and not by addresses in the packet.

REFERENCES

1. Davies, D. W., Holler, E., Jensen, E. D., Kimbleton, S. R., Lampson, B. W., LeLann, G., Thurber, K. J., and Watson, R. W. (1981). *Distributed Systems—Architecture and Implementation.* by Springer-Verlag, Berlin Heidelberg, New York, p. 486.

2. Deasington, R. (1982). *A Practical Guide to Computer Communications and Networking.* Published by Ellis Horwood Limited, England. Distributed by Halsted Press, a division of John Wiley and Sons, New York, p. 29.

8

Factory Information System Implementation

8.1 INTRODUCTION

In Chapter 2 the problem of accessing the manufacturing operation to conduct a needs analysis was explained. One of the approaches was to discover the needs by first designing a small simple factory information system (FIS) which is easy to modify and then support its natural evolution as users create new applications. This is also the best way to implement the FIS if the inherent slow rate of installation and use can be tolerated. Usually, the decision to automate a portion of a factory and therefore to implement a FIS is made on a larger scale. When this happens the project is large by definition, requiring significant planning and aggressive management. This chapter deals with the latter case.

Whether the FIS be simple or large and encompassing, a definition of what it should do in the long term is essential before

1. Needs Analysis
2. Justification I
3. Application Simulation
4. Application Specification
5. Justification II
6. System Design and Configuration
7. Justification III
8. Development System Installation*
9. Software Development
10. Target System Installation
11. Engineering Trials
12. User Procedures Development
13. Production Trials and Training
14. Impact Assessment
15. System Evolution

*May be omitted

Figure 8.1 Typical factory information system development phases.

design and implementation begins. The lack of this definition is always a major problem. Many manufacturing operations do not yet have the extra talent to address the definition of what the FIS should actually do.

A large factory information system (FIS) is usually implemented by a series of progressive phases. Many of the early phases end with the decision to stop or move ahead based on the evaluation of whether the usefulness of the FIS justifies its cost as perceived from knowledge available at that time. A list of these phases appears in Figure 8.1.

It is very advantageous to execute the early phases carefully and in detail. Occasionally, systems are implemented without going through all 15 of these phases. This usually occurs when the driving force for creating the FIS is not its end user but a supporting engineering or staff organization. If this happens the first phase, Needs Analysis, is sometimes minimized because those advocating the system feel they know what is needed. System development then really begins at phase 6, System Design and Configuration without the application being adequately defined. Now the first justification occurs at phase 7 instead of phase 2. If this justification is accepted and moneys are committed to hardware, the software development expenses begin. These software expenses are the major cost of most custom systems. From now on it gets continuously more difficult to stop the development program because the first time the user community has a chance to participate is during Procedures Development, (phase 12) after most of the system is operational.

If the FIS proves not to be as useful as anticipated and the investment has already been made, effort must then be expended on modifying the system to make it useful enough to justify its cost. This effort of course adds to the cost, making it increasingly difficult to reach break-even. The modification process usually starts with a belated, more thorough needs analysis, and so the system implementors are back at the beginning with considerably less credibility and funds. Hardware may have been purchased which is no longer needed because different hardware is now required. Application software, and in some cases even system and data base software, must be rewritten to meet the newly determined needs.

A much more effective approach is to force the FIS development to pass through each of the 15 phases with early emphasis on establishing exactly what the system will do to justify its cost. There must be a willingness to stop whenever it appears that the payback will not occur.

A good way to evaluate the potential impact of an FIS is to do the full scale needs analysis. The needs analysis can be justified with what Robert Schaffer calls the breakthrough project, which is described in Schaffer (1980). The breakthrough project is a small straightforward application, limited in scope, with measurable short-term goals that can be completed in six months or less. The results must be quantifiable and in concert with the user's business objectives. If this small project is successful, a broader full-scale needs analysis can usually be justified.

The rest of this chapter defines and reviews in greater detail each of the 15 phases listed in Figure 8.1 for a large factory information system development.

8.2   THE PHASES OF FIS DEVELOPMENT

8.2.1   Needs Analysis

Needs analysis, which was explained in greater detail in Chapter 2, is the investigation of what the manufacturing organization requires to significantly improve its performance. The scope of the analysis should be quite broad. Only a subset of the needs will probably be addressed by the FIS. The broader approach is required to establish priority among the many sets of needs that usually emerge during the analysis. Emphasis is placed on determining what is limiting high quality productivity. Often only some of these issues can be resolved by a FIS. The other issues, once defined, become objectives for management action aside from the FIS project.

The needs analysis is usually done by a team of people. The composition of this team is very important. Industrial engineers from within the manufacturing operation who are creative and have in the past made contributions to productivity, are respected by their peers, and work well with others are obvious choices. A leading production supervisor or manager should also participate to ensure the definition of management needs. Other desirable members are people who are not directly associated with the target manufacturing operation but have related experience in similar operations. These people might be internal consultants from the company research facility, outside consultants, or someone from a different department, plant, or division where similar manufacturing occurs. A third type of person are those who bring specific skills needed to implement the system. These can be system engineers, statisticians, software engineers, operations research specialists, industrial psychologists, etc. It is also wise to include form the onset a specific person from the financial staff to participate in setting measurable goals that will quantify the project's success.

A very important person on the needs analysis team is the unique, enthusiastic user. This person becomes the driving force behind much of the work required for the analysis and is essential for the successful later development of procedures for system use in phase 12. He or she is usually a leader within the user community who has some understanding and interest in computer systems and recognizes the opportunity afforded by association with the FIS project.

The needs analysis team defines the specific FIS objectives. These objectives should be reviewed by an executive of the user community who has the authority to make major decisions. The review should take place with the entire team present. This forces agreement and understanding by all on the key objectives. The meeting establishes a direction which will maintain the project's continuity through a number of organization changes for as long as three years. Following agreement on objectives, project tasks and responsibilities can be specified and timing can be planned with major milestone documentation.

The group then works with the manufacturing organization, spending time at the plant site to learn how the manufacturing operation currently functions, what its needs are, whether the needs can be met, and what it would cost in very approximate terms to meet various levels of needs.

A very important phenomenon can occur during this first phase that will increase system utilization and impact. The users of the system who participate in the needs analysis gain a sense of system ownership. This translates to an understanding of the objectives and how the final system is supposed to work. This user participation promotes a later eagerness to use the factory information system effectively.

8.2.2 Justification I

The objective of this justification is to determine whether there is a reasonable probability that the time and cost of meeting various levels of needs is a good investment. This means the tangible and intangible savings produced by fulfilling each need and combination of needs has to be approximated.

This is far from an exercise! Even if a quantitative value can not be assigned to each need or combination of needs the specific definition of what is gained when each need is fulfilled can place the needs into categories of valuable, helpful, or convenient to have. The needs analysis team, at this point, must recommend project termination if few needs have been determined or if most of them are only helpful or convenient. If a number of satisfied needs make valuable contributions and the cost of satisfying them is reasonable, movement to the next phase is appropriate. A second meeting with upper management to review the results of the needs analysis is recommended to formalize this transition.

Any expenditure for a FIS is competing with other proposals
that also require funds for implementation. Justification of an FIS
in terms that allow management to optimize the use of these funds
is reasonable and essential for good business practice. Quantify
each justification phase as much as possible to help with this decision process. Project costs will remain low until after the third
justification at phase 7.

8.2.3 Application Simulation

Simulation of the FIS can be used as a technique to confirm the
users judgement that a particular analysis or report is going to be
of value. If the simulation can be done using actual factory data,
many important practical problems are also revealed prior to defining system hardware. The technique is to collect the data manually or with some simple data recording device, analyze the data
using one of many commercial statistical and graphical software
programs such as those cited in references 2, 3, or 4, and produce
a report or graph of the results. This is a mock-up of what the report would be like if it were produced by the FIS.

The process of data collection, analysis, and mock-up reporting should continue for a few natural reporting cycles. For example, if the analysis is looking for statistically significant trends,
the simulation should cover enough time to test whether trends
are occurring. When a trend is identified the user will fully appreciate the potential of the FIS and its justification becomes much
more quantitative and concise. If a trend is not identified in the
expected time period, the assumption regarding the value of the
analysis has to be re-examined. It is much better to verify expected FIS impact at this phase of system development than it is to
try to create impact after a large system investment has been
made.

The simulation of a FIS using real factory data can produce
results that are identical to those of the final system except they
are produced slowly and are therefore not timely. The value of the
FIS can be quite accurately estimated, however, by reviewing with
the manufacturing organization what they would have done if the
report produced using a simulated FIS had been available earlier.
This leads to a determination for the system designers of how rap-

never finally established and the user, like someone selecting from
essential for system design and significantly influences the configuration phase which is still two steps ahead in our progression.

8.2.4   Application Specification

By now the level of worth of the system is established in the
minds of the team participating in the needs analysis but they
need to convey this quantitatively to the financial community and
in specific detail to the computer systems specialists or vendors.
The first step is to prepare an application specification. Techniques for doing this were discussed in Chapter 3. This specification defines exactly what the FIS will do in terms that all parties
can understand and interpret. This hard document is used many
ways. First, it lists the specific needs to be fulfilled. The impact of
these can therefore be estimated and summed for the next phase
of development.

The specification establishes the design objectives for hardware and software. It permits the computer systems engineers to
organize their phase of the FIS project into design and programming tasks, many of which can be done in parallel because the
total system needs have been defined up front rather than evolved
by experimentation. The specification also becomes the basic document for training the users who did not participate in the needs
analysis or system definition phases.

Finally, the specification establishes at this time in the project a set of fixed objectives that can be used to stop the propen-
sity to continually make changes which delay the beginning of
system use. From now on the cost of changing what the system
should do raises rapidly. Without some document defining specific
objectives it is easy to drift into a state of continuous system evaluation and change.

8.2.5   Justification II

The objective of this second justification is to more precisely estimate the system payback now that each specific solution to a need
has been determined and documented in the applications specification. Each need that can be met contributes to this payback.

There can also be serendipitous worth from combinations of fulfilled needs.

8.2.6   System Design and Configuration

System design and configuration is the task of specifically defining each component of hardware and software required to meet the application objectives. Often the same need can be met with different configurations. To further confound the problem, the choice of hardware and configuration, (how the various pieces of hardware are to be connected together), is also dependent on a number of other considerations beside the application per se. These can be, for example: reliability of system operation; response time to users; the number of users; timeliness of automatic data collection; physical environment; availability of maintenance support for the system; required time constraints for system recovery after a failure; and so forth.

The execution of this phase requires experienced, professional people with a depth of understanding in computer systems science. Decisions at this stage will have significant financial impact. Since the investment is still only engineering expense, rather than capital and software development, it pays to do this task carefully.

8.2.7   Justification III

This third and final evaluation before implementation firmly involves the management of the various organizations participating in the FIS venture. As a group they must either commit to support the project or stop further work before the rate of expenditure radically increases. If the project continues it is essential that the commitment be to project completion. A partially completed system is proportionally worth much less than the part completed because it must be augmented by continuing use of the old information systems to supply a complete picture of factory operations.

The commitment must also include accurate scheduling of human resources. Extreme frustration and very poor resource utilization occurs when one part of the project team falls behind schedule, idling other members. Poor project management, resulting in the unsynchronized application of the various skills, can increase the expense of implementation.

A favored approach to this justification is to configure a number of systems to meet various levels of needs and determine the approximate cost of each including the cost of software development, system installation, training, and documentation required for use. The value of having previously determined the worth of fulfilling each need and combination of needs now becomes apparent. In combination, ordered by estimated impact, each of these configurations have different capital and software costs. The optimum compromise is thus determined by contrasting the payback from each need and combination of needs against the corresponding system cost.

Ordering the comparison in descending impact usually results in finding that a minimum system cost exists which raises in steps as major system components and software are added. The anticipated impact of savings, however, starts fairly high but increases more and more slowly as the system gets bigger. A natural compromise is on a cost plateau where going to the next level of complexity and cost does not in itself provide a justifiable return on investment.

Participation by the executive management in the final review of this justification is essential. The original objectives can be reviewed and compared to what now appears feasible to do. Compatibility with current strategic plans is re-established. The new set of objectives derived from this review become the criteria for success. In many cases these are significantly different from the original and the executive managers need to be aware of these differences. The decisions made should be documented so they become institutionalized. This means the organization rather than any specific individuals has endorsed the project. Institutionalization is necessary because in many cases individual supporting managers will be reassigned or in some cases leave the company prior to project completion. Changes imposed by new management can be very disruptive in later phases and are not cost effective if the project worth has been properly determined.

8.2.8 Development System Implementation

It is necessary to set up an activity where the hardware used by the system can be interconnected and tested, and where the software can be developed and evaluated with the hardware. The best

place to do this is at the target manufacturing site. Locating this
activity away from the factory tends to separate it from the realities of the production operations. In some cases it must be at a
different location to save cost because the computer systems engineers and programmers are not physically located at the plant
site.

Another consideration is the need to keep this development
system intact and separate from the production system after the
latter is installed. This situation arises if major maintenance and
evolution of the FIS must be done remotely because of skilled
personnel location or because the nature of the manufacturing operation requires dedicated hardware. These situations occur frequently when the manufacturing location is outside of the home
country.

If the computer engineering team is resident at the manufacturing site, the development system may still not become the production system when implementation is completed. In some
applications up-time and reliability of operation dictates a separate
backup system. The development system can then fulfill this role
as well as be used for application evaluation, software maintenance, and noncritical general computing for the plant.

8.2.9 Software Development

The software required to run the FIS and to perform the applications defined by the application specification is produced during
this phase. Hardware selection and configuration should be strongly influenced by available software to minimize cost. There is no
need today to write operating system, data base management, or
communications software. Ideally, only the customized application programs need to be written. In the case of turn-key systems
even the application programs are furnished by the system vendor
or can be produced directly by the user without the services of
software specialists.

The software engineer must do extensive planning to design
the broad structure in which the application programs and data
files will function. If the application specification is already structured as described in Chapter 3, some of this planning has been
done and the production of software is easier and faster. This is
because the people who determine what a system will do have in-

sight into where it will most likely evolve. They therefore conscientiously structure the application specification for the requisite growth. This structuring for growth is particularly important when configuring the data base where trade-offs must be made between flexible growth and response time.

The schedule for software implementation should be as realistic as possible. Human resources are scheduled during this and the following three phases: to conduct engineering trials, to detect software problems and debug them, and to develop the startup procedures for system use. If software production falls behind schedule, the resources to do this testing will be prematurely available and thus inefficiently utilized. The final debugging of the software must occur on the production system. Inaccurate scheduling of this operation can have the further effect of disrupting manufacturing.

8.2.10 Target System Installation

The target (production) system hardware can be installed while software development proceeds on the development system if one is available. If the development system is to be used for manufacturing its installation should be planned to include as much flexibility as possible. Changes inevitably become desirable as a result of the software development and testing. Flexibility permits these to be incorporated with minimum expense and program delay.

A number of serious technical and environmental issues are related to the target installation. These are listed below and discussed in the following paragraphs.

1. Ambient temperature

2. Dust

3. Noise level

4. Fire protection

5. Security

6. Maintenance

7. Power source

Ambient Temperature

Solid state devices used in computers are more reliable and therefore the cost of maintenance is less if the equipment is well
cooled. The computer and its peripheral equipment should be in a
cold environment. This is usually a separate room with independent air conditioning. Computer equipment vendors will provide
guidance but they tend to minimize the thermal problem. It is best
for the user to err on the cold side. Do not plan to use the computer room as space for constant human attendance. It will be too
cold if properly maintained.

Dust

Particulate matter is the great destroyer of disks and magnetic tape
stations. It also blocks the flow of cooling air through filters. Dust
will clog the filters which are necessary in all systems to keep dirt
away from record and read heads. This reduces air flow and causes
thermally induced failures irrespective of incoming air temperature. Dust in disks and tape stations will irrevocably destroy the
stored data and the storage device by physically abrading the magnetically alterable recording surface and the heads themselves. As
computing technology evolves, the physical spacings are becoming
smaller, making dust even more of a threat. To avoid these problems the incoming air should be filtered prior to entering the room
or plenum used as a source for cooling. Human traffic in the computer room can also be a source of dust and should be kept at a
minimum.

Noise Level

Noise level is not the problem that it used to be, but disk drive and
impact printer noise when coupled with the cold temperature certainly make the typical computer room an objectional place to
live. Quiet operation is one of the many advantages of laser
printers.

Fire Protection

Water and electrical equipment don't mix. Permanent damage to
computing equipment can occur from a sprinkler system installed
in or over the installation. Self contained systems for automatically flooding the computer room with extinguishing inert gas are

available and recommended. Personnel must be trained to avoid the danger of asphyxiation by leaving the room immediately if the extinguisher is discharged.

Security

As industrial dependency on computer-based systems grows the value of their contents and the effect of losing their operational capability becomes greater. The incidence of unauthorized acquisition of sensitive data is becoming a significant problem. The malicious or unintentional destruction of historical data, operating system capability, or hardware will increase as people lose their fear of computing systems and become more adept at their subversion. Three forms of security should be considered: (1) limit physical intrusion into the computer room, (2) permit only legitimate local users electronic access to the system, and (3) institute protection from access by external persons through communication links to other systems or via telephone modems. Most systems provide multiple level pass word protection. An access monitor is desirable so the system administrator can periodically identify who has entered and used the system.

Maintenance

If an FIS system is to be useful it has to run reliably for long, continuous periods of time (weeks) and when it fails the repair time has to be a matter of hours, not days. A new installation should start-up with a vendor maintenance contract defining acceptable response. In-house maintenance by the user organization can be the most timely and cost effective but only after the required technical expertise, experience, and stock of replacement parts have been acquired.

Power Source

Everything that can be imagined will happen to the power source in accordance with Murphy's more pessimistic laws. Noise and cycle drops which can initiate false signals in the computer system can be avoided by using capacitor based power conditioners. Longer outages and brown-outs require battery backup, the intelligent selection of nonvolatile memory, and perhaps a motor generator to automatically start before the battery power runs out. The degree of protection depends on the incidence of power failure and the seriousness of the consequences. Personnel at the target site usually have a good idea of the magnitude of the power problem only if other similar computers are already on location. If a power problem exists it should be addressed during phase 6 while the system design and configuration is taking place. Don't wait until after the hardware delivery to realize that the manufacturing site's power sources are not reliable enough to support a system that loses its value for hours every time the power fails for a few seconds!

8.2.11 Engineering Trials

The engineering trials are tests of the system hardware, software, and applicability. These trials are usually conducted by the team of people who have worked on the system since the beginning of the needs analysis. It is premature for these tests to be done by the user. So many problems develop during the trials that unless the users have prior experience with system start-ups they could easily become disenchanted with their FIS before having the chance to use it effectively. It is therefore appropriate that the engineering trials and the solution of resulting problems be conducted by the engineering staff prior to user involvement.

Most FISs provide a series of application programs that collect data from manufacturing processes, subject the data to analysis, and produce reports and control directives that help direct operations. Each one of these should be tested. Often the solution to a problem with one application program or data collection program will affect other programs. For this reason it is not wise during trials to approve some applications for use while others remain untested. After the engineering trials have verified the operational integrity of the system, the development of procedures for system use can begin.

8.2.12 User Procedures Development

The people who did the needs analysis possess a good concept of how the FIS should be used. It was this concept that drove the application specification. The application specification tells what the system is to do. It does not describe how the system is accessed or how the output from the system is used to control manufacturing operations. Procedure development addresses these issues of how and when the user accesses the system and more importantly, what actions the user takes in response to system output. The concept of how to use the system invariably changes as the system is implemented. In some cases, a couple of years may have elapsed between when the needs analysis was done and when the system becomes technically operational.

The procedures developed during this phase define exactly how the start-up system will be used. This task is done jointly by the needs analysis team and the initial users of the system. The process begins as soon as the engineering trials are complete and continues through the production trials and training phase.

Unlike the engineering trials, this phase must have a high level of user participation. Often the participants of the user community have changed since the needs were defined. New users, by participating in the procedures development can learn what was previously defined as needs and make their contributions by improving the way the system is applied. Active user participation provides a cadre of experienced people within the user community who can act as trainers for the rest. Since these people have had the opportunity to contribute to the procedures they are usually highly motivated to teach during the training phase. The users who participate need to be experienced, creative, and have good communication skills. They are a group key to the successful transfer of the system from implementors to users. The best production personnel should be selected rather than settling for those who have the time to participate because they occupy less responsible positions.

The most important aspect of procedures development is its influence on rapidly getting a new system to be productive. Many systems languish, after they become operational, because the developers leave procedures development to the users expecting them to recognize and address this need. Experience dictates that the user has many more urgent tasks competing for attention. Most of the users will not know what the system does. Even those who do must spend an inordinate amount of unavailable time if they are to define the procedures without help from the needs analysis team.

Procedures for use are usually derived for each level of the production organization and for each natural reporting time period. For example the superintendent may retrieve certain performance reports at the beginning of each shift and react to their contents by issuing directions for actions to supervisors. Each hour during the shift, the supervisors may look at another set of reports and direct more responsive actions such as on-line maintenance. Process technicians may react immediately to real time out-of-control conditions signaled by the system. Separate procedures need to be defined ahead of time for how each level of user reacts.

A simple way to document procedures is by using a logic flow diagram. Branch points for various actions based on system output can be concisely shown in summary form. After the validity of the procedure is proven, the flow diagram can be implemented in software resulting eventually in a system which provides suggested directives based on management-by-exception reporting and procedures logic. Thus, well documented procedures, in addition to getting the FIS used and thereby initiating payback, set the stage for evolution to a higher level of control through faster automated decision processing. It is at this level that the use of artificial intelligence (AI) programs come into play.

8.2.13   Production Trials and Training

This phase is characterized by intensive formal user training plus production use to solidify the procedures. A well publicized beginning and ending as well as a serious level of formality stimulates the users to put effort into the learning process. The serious nature of this phase also impresses upon them management's dedication to the system as a major new tool with incentive ramifications. The beginning of this phase will be recognized by the plant management as the end of system development. Upper management will mark the end of this phase as the start of realizing the promised return on investment as performance begins to improve.

An effective way to execute the trials is to start up gradually, one logical group of application programs at a time. The needs analysis usually defines groups of applications that address specific problems. These groups are sufficiently independent to be started up in series. This technique is almost the only way to have enough resources for adequate training. After all the applications are running, in-depth training on the interactions between applications is usually in order.

Most of the FISs seem to falter initially or are significantly delayed in their application for two reasons. The first reason is that incentives for use are not forcefully portrayed by the top plant management. The system should not be committed to production until the management is convinced it will improve performance. With this conviction must come a very clear directive that it is to be used to the exclusion of past methods of control.

The second reason is that user training is minimized or neglected. Training takes time away from production. The best time for this training to occur is when the economic cycle is down. During such a period of reduced production pressure, people can learn how to use the system before the next expansion period begins. Production personnel always seem able to use these systems effectively once they learn how. They can not take advantage of this tool if they do not know what it provides, how to get at this information, and what to do with it.

Plant management must think of this training as transcending just the FIS. The entire manufacturing process was examined to determine what the system should do and how the users should react to system output. The training therefore needs to include a review of the manufacturing process. Many times this review uncovers a lack of basic process knowledge that can be eliminated as part of the instruction for system access and operating procedures. The FIS training often has to include the basic process to make procedures relevant. The typical user's reaction to this more encompassing training has been relief from the anxiety of dealing with unknowns and renewed interest from the gain in fundamental understanding. Motivation through understanding builds high morale and productivity.

Many FIS installations require specialized courses in process control and statistics. An understanding of statistical control fundamentals and techniques is required for intelligent system application. Almost all modern systems profitably use statistical tests to initiate management-by-exception reports and statistical control for processes and equipment maintenance.

8.2.14 Impact Assessment

What has the FIS actually accomplished? How are these accomplishments related to the initial goals? Has the gain in productivity been worth the investment in the system? What are the expected future gains? What should be done differently as the system evolves or as other systems are considered. The answers to these questions are not easily quantified but they are necessary to improve our manufacturing capability.

The impact assessment has to be done in stages as follows:

1. Have the financial member of the needs analysis team help quantify the objectives in measurable terms as the needs analysis progresses.

2. Establish a manufacturing performance baseline prior to system implementation. This then becomes the "before" of the before-and-after comparison.

3. The third step is to immediately, after the system becomes operational, establish a second baseline using performance parameters now collected and analyzed by the system itself.

4. The fourth step is to measure the change from these baselines and quantify them. This last step usually can not be done until about a year after the system has been operating in the production mode. A common mistake is to measure impact prematurely before the system is being used effectively. The results are understandably poor and can lead to labeling the project a failure when in actuality, a year later it is very successful. An empirical rule is that it takes as long to get a system used effectively as it took in its development and implementation.

8.2.15 System Evolution

The good system is one that can evolve. A poor system lacks this flexible quality. As the manufacturing process and mix of product change, the system must be able to adapt or it will grow obsolete. Evolution capability is also necessary for optimum impact. The best needs analysis and application simulations always omit some aspects that enhance usefulness. These additional needs become apparent as the system is used. An evolutionary phase which continues until eventual obsolescence can and should be anticipated.

8.3 CONDITIONS AND STRATEGIES FOR SUCCESSFUL IMPLEMENTATION

In conclusion, there are a number of conditions that should be considered to ensure the success of a FIS implementation. In this context implement means the entire process from conception of the need for a FIS to the users consistent use of the installed system. To be considered successful, a system has to be designed, built, and installed within agreed to time-budget constraints and the system must perform according to user expectations.

The conditions offered for consideration are:

1. A need exists

2. The need is recognized by the user

3. Management is supportive

4. Required resources are available

5. Training actually occurs

6. User interfaces are simple

7. There is management dedication to an evolving system

8. The project team is result oriented

9. Interfaces to other systems are provided

10. User expectations are minimal

11. System start-up is phased

8.3.1 Need Exists

It may be that the impetus for the FIS is not a real need but a perceived need. Sometimes it can stem from an emotional desire to appear up-to-date and modern. For example there are many robotic installations which are not needed but were implemented simply so the factory could advertise the use of robots. They are usually justified under the guise of learning about a new technology.

To assure that real need exists requires an analysis of how the factory is controlled. To do this one must know what data is collected, how the data is analyzed, how the results of the analysis are presented, and how the results of the analysis are used and by whom. If better techniques are apparent from the use of a FIS, then the impact of using these should be estimated. If significant impact results, then a real need has been validated.

8.3.2 The Need Is Recognized by the User

The best way to get users to see the need is to involve them in the needs analysis and system design process. Initially, this is done informally in meetings where ideas are generated and exchanged. The results of these meetings then become the more formal written proposal and presentations to higher levels of management.

Hopefully, as a result of these meetings, enthusiastic users are identified. These are people who understand the manufacturing operation, have the confidence of the plant management, know some of the technology of computer systems, and are creative. Such people can then help with the simulation of analyses and reports to show other users the potential value of the system.

8.3.3 Management Support

Without executive support funding will not become available. The Chief Executive Officer or Vice President should therefore participate in the planning reviews.

8.3.4 Required Resources Are Available

The project management team has to be defined. Various kinds of people need to be available. These include the enthusiastic users, people to specify the applications to be implemented, hardware and software design engineers, and training personnel for all levels of people who will use the system. Obviously, along with these people must come the money for hardware and site conditioning.

8.3.5 Training Actually Occurs

A training program should begin before the FIS is made operational and should be a continuing effort as the system and people who use it change with time. Unfortunately, in over half of the installations, training is talked about but not implemented. As a result, the FIS is poorly used and the expected return is not realized.

8.3.6 User Interfaces Are Simple

The labor associated with manual input of data and system commands has to be minimized by the use of simple, self-prompting, input-output devices that require very few key strokes or motions. The major portion of data should be input automatically by production machines and monitors. Analysis and report generation should be initiated by exception whenever possible to reduce the effort required to search through awesome reports to detect a problem and obtain information for its solution. The operation and maintenance of the FIS should also be made routine so that the user can do these tasks without the help of professional computer personnel.

8.3.7 Dedication to an Evolving System

Good FISs change as the needs change. Poor systems become obsolete. The management of the plant should support, with people and hardware, long term evolutionary change to keep the FIS from becoming obsolete. Change will occur in the application programs and in the system configuration.

8.3.8 Results Oriented Project Team

Computer people enjoy building systems. The objective of an FIS implementation is, however, its useful impact. It is not to provide work for software and system engineers. The user should therefore participate and agree in detail that the system functions being implemented are justified.

8.3.9 Provide Interfaces to Other Systems

Computer aided design, financial, strategic sourcing, production control and scheduling, and the factory office are some of the interfaces that should be considered during implementation.

8.3.10 Minimize User Expectations

All humans read into situations what they want to hear and see. The potential user of an FIS is no exception. Astronomical expectations can easily grow from innocent remarks. If this occurs there is no way to prevent disappointment when the system is made operational. Frequently the user hopes the system will solve all his problems because it is a computer! The implementor can detect this failing. It manifests itself as the user's inability or unwillingness to define specifically what the system is to do. This means both of you are in trouble from the very beginning of the project.

8.3.11 Phase System Start-up

Small groups of applications have to be started because there are not enough trained personnel to start everything at once. The next group of applications should not be started until users are confident and proficient.

REFERENCES

1. Schaffer, R. H. (1980). "Productivity Breakthroughs: Begin With Results Not Preparations", Manufacturing Productivity II Conference. Obtain from: Robert H. Schaffer and Associates, 401 Rockrimmon Road, Stamford, CT. 06903.

2. Minitab Project, Statistics Department, The Pennsylvania State University, University Park, Pa. 16802

3. RS/1 Statistical and Graphical Analysis package from Bolt Beranek and Newman Inc. (BBN Sortware Products Corp.), Cambridge, MA 02238

4. BHARAT Intelligent Graphics package from Siemens Research and Technology Laboratories, Princeton, NJ. 08540

Index

Please note: Italic page numbers indicate main headings or major information about the entry.

9 780367 451455